Applied Mathematical Sciences

Founding Editors

F. John
J. P. LaSalle
L. Sirovich

Volume 215

The mathematization of all sciences, the fading of traditional scientific boundaries, the impact of computer technology, the growing importance of computer modeling and the necessity of scientific planning all create the need both in education and research for books that are introductory to and abreast of these developments. The purpose of this series is to provide such books, suitable for the user of mathematics, the mathematician interested in applications, and the student scientist. In particular, this series will provide an outlet for topics of immediate interest because of the novelty of its treatment of an application or of mathematics being applied or lying close to applications. These books should be accessible to readers versed in mathematics or science and engineering, and will feature a lively tutorial style, a focus on topics of current interest, and present clear exposition of broad appeal. A compliment to the Applied Mathematical Sciences series is the Texts in Applied Mathematics series, which publishes textbooks suitable for advanced undergraduate and beginning graduate courses.

Daniela Calvetti · Erkki Somersalo

Bayesian Scientific
Computing

 Springer

Daniela Calvetti
Department of Mathematics, Applied
Mathematics, and Statistics
Case Western Reserve University
Cleveland, OH, USA

Erkki Somersalo
Department of Mathematics, Applied
Mathematics, and Statistics
Case Western Reserve University
Cleveland, OH, USA

ISSN 0066-5452 ISSN 2196-968X (electronic)
Applied Mathematical Sciences
ISBN 978-3-031-23826-0 ISBN 978-3-031-23824-6 (eBook)
https://doi.org/10.1007/978-3-031-23824-6

Mathematics Subject Classification: 65C40, 65F22, 65F08, 62C10, 62F15

This Springer imprint is published by the registered company Springer Nature Switzerland AG
The registered company address is: Gewerbestrasse 11, 6330 Cham, Switzerland

To Oriana and Giovanni Madan

Preface

Fifteen years ago, when the idea of using probability to model unknown parameters to be estimated computationally was a less commonly accepted idea than it is today, writing a book to show how natural and rather straightforward it was to merge Bayesian inference and scientific computing felt a brave and daring act. The natural environment for such a symbiotic relation to develop was inverse problems, a well-established research area where both Bayesian inference and computational methods were establishing their own separate strongholds. One of the main reasons for the separation is that probability was not part of the usual training of a numerical analyst, just like scientific computing was not part of the typical background of a statistician or a probabilist. Nonetheless, the concurrent similarity and complementarity of the two fields suggested that if there could be a way to create some synergy between them, it would probably lead to novel and unexpected findings.

We ourselves, coming from numerical analysis and Bayesian-physics-based inverse problems, were very aware of some of the obstacles to be overcome, including having to fill some holes in the background without getting too deep into the intricacies of these fields and feeling like we had ventured too far out of our own comfort zone. The book *Introduction to Bayesian Scientific Computing: Ten Lectures in Subjective Computing* [18] was a light-hearted gateway to what we called Bayesian Scientific Computing, where we deliberately chose to not include theorems or proofs, emphasizing instead the problem-solving power of the approach. The book was written with the purpose of awaking the curiosity of unlikely readers, and in a style that, we hoped, would make the reading experience more like following an unfolding story rather than scientific reading. Likewise, when it came to compiling a list of references, we went for a quite radical version of Occam's razor. In hindsight, our first book on Bayesian scientific computing reflected our enthusiasm for the content and our desire to make it easy for newcomers to look into it, preemptively answering many of the questions that the reader may have had and would not dare to ask, in fear of sounding naïve. Over the years, as the idea of combining classical numerical analysis and Bayesian inference became more mainstream, the questions that people asked became deeper, and we started to feel about our book the same way as we did about the bell-bottom pants and the mullet haircuts of our younger years.

In the last 15 years, pairing Bayesian inference and scientific computing has become more natural, also thanks to the tremendous growth of uncertainty quantification, an area of applied mathematics where Bayesian inference and scientific computing fit very naturally. Over the same period of time, Bayesian scientific computing has advanced a lot, and we began to feel that it was time to update the 2007 book and write a more grown-up version that would include the new developments in the field. That would also give us a chance to revisit some of the topics that we, in our initial enthusiasm, had sometimes treated in a very cavalier way. Among the topics that were either missing, or not systematically developed in the 2007 monograph, are hierarchical models and sparsity promotion priors. Sparse solutions of inverse problems, and methods for computing them, have been the topic of active research in the last 15 years, in particular, following the popularity of ℓ_1 and total variation regularization. Since sparsity is part of the a priori belief about the unknown, it is very natural to express it in Bayesian terms. The systematic analysis of sparsity-promoting priors carried out in the last 15 years within the larger family of hierarchical priors has shed light on how to characterize sparsity in probabilistic terms. Hierarchical priors and their computationally friendly formulations, examined at different levels and from various points of view over several chapters, are prominently featured in this monograph, and the connection between certain choices of hyperpriors and Tikhonov regularization is also established. Another major topic that was completely omitted from [18] is sequential Bayesian methods, which now occupy the last two chapters of the book. Particle filters, Kalman filters and Ensemble Kalman filters combine, in an elegant and organic way, many of the Bayesian scientific computing topics introduced in the earlier chapters. The evolution and observation equations can be interpreted as prior and likelihood, and the way in which the posterior is updated from one time step to the next resembles the natural learning process. Likewise, sampling methods have a more prominent position, commensurate with their key role in many Bayesian scientific computing tasks. The rational for Monte Carlo integration in high dimensions is an excellent example of the need for sampling from a distribution, and is used here as a gateway to some of the most popular sampling schemes, including the intuitive importance sampling, the classical Gibbs sampler and Metropolis–Hastings algorithms, all the way to the clever preconditioned Crank–Nicolson sampler.

Other significant ways in which this book differs from [18] are in the level of detail and notations. During the years, we have experienced that some of our shorthand notations in the previous book led to confusion and potential ambiguities, and in order to avoid that, some notations in the new book are arguably a bit heavier but less prone to misinterpretations. We have tried to retain some of the lightness of the previous book, and we hope that the style manages to elucidate the origins of the ideas and makes the book a rewarding reading experience, and not just a source of reference.

One thing between the two book projects is invariant: Like the old book, the new one is also the result of a dialogue with our students. The new book corresponds to the curriculum of the graduate course Bayesian Scientific Computing taught over the years at Case Western Reserve University, and much of the changes and additions are a result of the questions and comments of the smart and inquisitive CWRU students from all over the campus. Moreover, we have had the opportunity to teach parts of this material in several other universities and research institutes, including Instituto de Matemática Pura e Aplicada (IMPA) in Rio de Janeiro, Brazil in 2011; University of Copenhagen, Denmark in 2014; University of Naples "Federico II" in Naples, Italy in 2015; Basque Center for Applied Mathematics (BCAM) in Bilbao, Spain in 2015; University of Warwick in Warwick, UK in 2015; University of Rome "La Sapienza" in Rome, Italy in 2015; MOX at Milan Polytechnic University in Milan, Italy in 2019; Danish Technical University in Lyngby, Denmark in 2019, as well as Gran Sasso Science Institute (GSSI) in L'Aquila, Italy in 2021. Our warmest thanks go to Jorge Zubelli, Sami Brandt, Gerardo Toraldo, Salvatore Cuomo, Luca Gerardo Giorda, Andrew Stuart, Francesca Pitolli, Barbara Vantaggi, Simona Perotto, Per Christian Hansen and Nicola Guglielmi for making these visits possible, and especially to the fantastic groups of graduate students, postdocs and local researchers participating in those courses, whose enthusiasm convinced us that we should continue to develop the material and that refreshing the book was the right thing to do. We are also very grateful to Howard Elman, whose insightful comments on a near final draft of this book helped improve the presentation of some topics.

If many of the topic in this book have played a central role in our research projects, our research interests have had a strong influence on the structure of the book. We gratefully acknowledge partial support for the work of Daniela Calvetti by the grants NSF-DMS 1522334, NSF-DMS 1951446, Simons Fellowship in Mathematics "The inverse problem of magnetoencephalography," and of Erkki Somersalo by the grants NSF-DMS 1016183, NSF-DMS 1312424 and NSF-DMS 1714617.

Cleveland, USA   Daniela Calvetti
June 2022   Erkki Somersalo

Preface to the 2007 Book *Introduction to Bayesian Scientific Computing*

The book of nature, according to Galilei, is written in the language of mathematics. The nature of mathematics is being exact, and its exactness is underlined by the formalism used by mathematicians to write it. This formalism, characterized by theorems and proofs, and syncopated with occasional lemmas, remarks and corollaries, is so deeply ingrained that mathematicians feel uncomfortable when the pattern is broken, to the point of giving the impression that the attitude of mathematicians towards the way mathematics should be written is almost moralistic. There is a definition often quoted, "A mathematician is a person who proves theorems," and a similar, more alchemistic one, credited to Paul Erdös, but more likely going back to Alfréd Rényi, stating that "A mathematician is a machine that transforms coffee into theorems."[1] Therefore it seems to be the form, not the content, that characterizes mathematics, similarly to what happens in any formal moralistic code wherein form takes precedence over content.

This book is deliberately written in a very different manner, without a single theorem or proof. Since morality has its subjective component, to paraphrase Manuel Vasquez Montalban, we could call it *Ten Immoral Mathematical Recipes*.[2] Does the lack of theorems and proofs mean that the book is more inaccurate than traditional books of mathematics? Or is it possibly just a sign of lack of coffee? This is our first open question.

Exactness is an interesting concept. Italo Calvino, in his *Lezioni Americane*,[3] listed exactness as one of the values that he would have wanted to take along to the twenty-first century. Exactness, for Calvino, meant precise linguistic expression, but in a particular sense. To explain what he meant by exactness, he used a surprising example of exact expression: the poetry of Giacomo Leopardi, with all its ambiguities and suggestive images. According to Calvino, when obsessed with a formal language

---

[1] That said, academic mathematics departments should invest on high-quality coffee beans and decent coffee makers, in hope of better theorems. As Paul Turán, a third Hungarian mathematician, remarked, "weak coffee is fit only to produce lemmas."

[2] M. V. Montalban: *Ricette immorali* (orig. *Las recetas inmorales*, 1981), Feltrinelli, 1992.

[3] I. Calvino: *Lezioni Americane*, Oscar Mondadori, 1988.

that is void of ambiguities, one loses the capability of expressing emotions exactly, while by liberating the language and making it vague, one creates space for the most exact of all expressions, poetry. Thus, the exactness of expression is beyond the language. We feel the same way about mathematics.

Mathematics is a wonderful tool to express liberally such concepts as qualitative subjective beliefs, but by trying to formalize too strictly how to express them, we may end up creating beautiful mathematics that has a life of its own, in its own academic environment, but which is completely estranged to what we initially set forth. The goal of this book is to show how to solve problems instead of proving theorems. This mischievous and somewhat provocative statement should be understood in the spirit of Peter Lax' comment in an interview given on the occasion of his receiving the 2005 Abel Prize[4]: "When a mathematician says he has solved the problem he means he knows the solution exists, that it's unique, but very often not much more." Going through mathematical proofs is a serious piece of work: we hope that reading this book feels less like work and more like a thought-provoking experience.

The statistical interpretation, and in particular the Bayesian point of view, plays a central role in this book. Why is it so important to emphasize the philosophical difference between statistical and non-statistical approaches to modeling and problem-solving? There are two compelling reasons.

The first one is very practical: admitting the lack of information by modeling the unknown parameters as random variables and encoding the nature of uncertainty into probability densities gives a great freedom to develop the models without having to worry too much about whether solutions exist or are unique. The solution in Bayesian statistics, in fact, is not a single value of the unknowns, but a probability distribution of possible values that always exists. Moreover, there are often pieces of qualitative information available that simply do not yield to classical methods, but which have a natural interpretation in the Bayesian framework.

It is often claimed, in particular by mathematician in inverse problems working with classical regularization methods, that the Bayesian approach is yet another way of introducing regularization into problems where the data are insufficient or of low quality, and that every prior can be replaced by an appropriately chosen penalty. Such statement may seem correct in particular cases when limited computational resources and lack of time force one to use the Bayesian techniques for finding a single value, typically the maximum a posteriori estimate, but in general the claim is wrong. The Bayesian framework, as we shall reiterate over and again in this book, can be used to produce particular estimators that coincide with classical regularized solutions, but the framework itself does not reduce to these solutions, and claiming so would be an abuse of syllogism.[5]

The second, more compelling reason for advocating the Bayesian approach, has to do with the interpretation of mathematical models. It is well understood, and generally accepted, that a computational model is always a simplification. As

---

[4] M. Raussen and C. Skau: *Interview with Peter D. Lax*. Notices of the AMS **53** (2006) 223–229.

[5] A classic example of similar abuse of logic can be found in elementary books of logic: while it is true that Aristotle is a Greek, it is not true that a Greek is Aristotle.

George E. P. Box noted, "all models are wrong, some are useful." As computational capabilities have grown, an urge to enrich existing models with new details has emerged. This is particularly true in areas like computational systems biology, where the new paradigm is to study the joint effect of huge number of details[6] rather than using the reductionist approach and seeking simplified lumped models whose behavior would be well understood. As a consequence, the computational models contain so many model parameters that hoping to determine them based on few observations is simply impossible. In the old paradigm, one could say that there are some values of the model parameters that correspond in an "optimal way" to what can be observed. The identifiability of a model by idealized data is a classic topic of research in applied mathematics. From the old paradigm, we have also inherited the faith in the power of single outputs. Given a simplistic electrophysiological model of the heart, a physician would want to see the simulated electrocardiogram. If the model was simple, for example, two rotating dipoles, that output would be about all the model could produce, and no big surprises were to be expected. Likewise, given a model for millions of neurons, the physician would want to see a simulated cerebral response to a stimulus. But here is the big difference: the complex model, unlike the simple dipole model, can produce a continuum of outputs corresponding to fictitious data, never measured by anybody in the past or the future. The validity of the model is assessed according to whether the simulated output corresponds to what the physician *expects*. While when modeling the heart by a few dipoles, a single simulated output could still make sense, in the second case the situation is much more complicated. Since the model is overparametrized, the system cannot be identified by available or even hypothetical data and it is possible to obtain completely different outputs simply by adjusting the parameters. This observation can lead researchers to state, in frustration, "well, you can make your model do whatever you want, so what's the point."[7] This sense of hopelessness is exactly what the Bayesian approach seeks to remove. Suppose that the values of the parameters in the complex model have been set so that the simulated output is completely in conflict with what the experts expect. The reaction to such output would be to think that the current settings of the parameters "must" be wrong, and there would usually be unanimous consensus about the incorrectness of the model prediction. This situation clearly demonstrates that some combinations of the parameter values have to be excluded, and the exclusion principle is based on the observed data (likelihood), or in lack thereof, the subjective belief of an expert (prior). Thanks to this exclusion, the model can no longer do whatever we want, yet we have not reduced its complexity and thereby its capacity to capture complex, unforeseen, but possible, phenomena. By following the principle of exclusion and subjective learned opinions, we effectively narrow down the probability distributions of the model parameters so that the model produced plausible results. This process

---

[6] This principle is often referred to as *emergence*, as new unforeseen and qualitatively different features emerge as a sum of its parts (cf. physics → chemistry → life → intelligence). Needless to say, this holistic principle is old, and can be traced back to ancient philosophers.

[7] We have actually heard this type of statement repeatedly from people who refuse to consider the Bayesian approach to problem-solving.

is cumulative: when new information arrives, old information is not rejected, as is often the case in the infamous "model fitting by parameter tweaking," but included as prior information. This mode of building models is not only Bayesian, but also Popperian[8] in the wide sense: data is used to *falsify* hypotheses thus leading to the removal of impossible events or to assigning them as unlikely, rather than to verify hypotheses, which is in itself a dubious project. As the classic philosophic argument goes, producing one white swan, or, for that matter, three, does not prove the theory that all swans are white. Unfortunately, deterministic models are often used in this way: one, or three, successful reconstructions are shown as proof of a concept.

The statistical nature of parameters in complex models serves also another purpose. When writing a complex model for the brain, for instance, we expect that the model is, at least to some extent, generic and representative, and thus capable of explaining not one but a whole population of brains. To our grace, or disgrace, not all brains are equal. Therefore, even without a reference to the subjective nature of information, a statistical model simply admits the diversity of those obscure objects of our modeling desires.

This book, which is based on notes for courses that we taught at Case Western Reserve University, Helsinki University of Technology and at the University of Udine, is a tutorial rather than an in-depth treatise in Bayesian statistics, scientific computing and inverse problems. When compiling the bibliography, we faced the difficult decision of what to include and what to leave out. Being at the crossroad of three mature branches of research, statistics, numerical analysis and inverse problems, we were faced with three vast horizons, as there were three times as many people whose contributions should have been acknowledged. Since compiling a comprehensive bibliography seemed a Herculean task, in the end, the Occam's razor won and we opted to list only the books that were suggested to our brave students, whom we thank for feedback and comments. We also want to thank Dario Fasino for his great hospitality during our visit to Udine, Rebecca Calvetti, Rachael Hageman and Rossana Occhipinti for help with proofreading. The financial support of the Finnish Cultural Foundation for Erkki Somersalo during the completion of the manuscript is gratefully acknowledged.

Cleveland, USA   Daniela Calvetti
Helsinki, Finland   Erkki Somersalo
May 2007

---

[8] See A. Tarantola: *Inverse problems, Popper and Bayes*, Nature Physics **2** 492–494, (2006).

Contents

Chapter 1
Bayesian Scientific Computing and Inverse Problems

Bayesian scientific computing, as understood in this text, is a field of applied mathematics that combines numerical analysis and traditional scientific computing to solve problems in science and engineering with the philosophy and language of Bayesian inference. An overarching goal is to apply these techniques to solve computational inverse problems by using probabilistic methods with efficient numerical tools.

The need for scientific computing to solve inverse problems has been recognized for a long time. The need for subjective probability, surprisingly, has been recognized even longer. One of the founding fathers of the Bayesian approach to inverse problems, in addition to reverend Thomas Bayes[1] himself, is Pierre-Simon Laplace[2], who used Bayesian techniques, among other things, to estimate the mass of the planet Saturn from astronomical observations.

The design of predictive mathematical models usually follows the principle of causality, meaning that the model predicts the consequences of known causes. The models are often local in the sense that changes in a quantity are described by the behavior of the quantity itself in a local neighborhood, the proper language, therefore, being differential calculus. In inverse problems, the goal is to infer on unknown causes when the consequences are observed. Inverse problems often lead to non-local models, involving typically integral equations. The classical Second Law of Thermodynamics states that in a closed system, entropy increases, which can be

---

[1] Thomas Bayes (1702–1761), in his remarkable *An Essay towards solving a Problem in the Doctrine of Chances* that was read in the Royal Society posthumously in 1763, first analyzed the question of inverse probability, that is, how to determine a probability of an event from observation of outcomes. This has now become a core question of modern science.

[2] Pierre-Simon Laplace (1749–1827): French mathematician and natural scientist, one of the developers of modern probability and statistics, among other achievements. His *Mémoire sur la probabilité des causes par les évènemens* from 1774 shows the Bayes' formula in action as we use it today.

D. Calvetti and E. Somersalo, *Bayesian Scientific Computing*, Applied Mathematical Sciences 215, https://doi.org/10.1007/978-3-031-23824-6_1

interpreted by saying that causal processes loose information. Therefore, inverse problems fight against the loss of information.

There is an almost perfect parallelism between inverse problems and Bayesian methods. The famous formula of Bayes is a tool to reverse the roles of two mutually dependent probabilistic events: If the probability of one is known assuming the other, what is the probability of the latter assuming the former? Arranging the events as causal ones, one arrives at the core question of inverse problems. However, there is more: The Bayesian framework gives tools to import additional information about the unknown complementary to what is included in the forward model, thus helping in the process of fighting the loss of information. Therefore, the Bayesian framework provides the perfect language for inverse problems.

The interplay of Bayesian methods and scientific computing is not limited to inverse problems. Numerical methods have long been used in statistics, in particular in what is known as computational statistics. One of our goals is to promote and nurture a flow in the opposite direction, showing how Bayesian statistics can provide a flexible and versatile framework for computational methods, in particular those used in the solution of inverse problems. This book addresses a collection of topics that are usually taught in numerical analysis or Bayesian statistics courses, with the common thread of looking at them from the point of view of retrieving information from indirect observations by effective numerical methods.

1.1 What Do We Talk About When We Talk About Random Variables?

Intuitively, *random events*, and the associated *probabilities*, have a well-defined and clear meaning[3] and yet have been the subject of extensive philosophical and technical discussions. Although it is not the purpose of this book to plunge too deeply into this subject matter, it is useful to review some of the central philosophical concepts and explain the point of view taken in the rest of the exposition.

A random event is often implicitly defined as the complement of a *deterministic event*, in the sense that if the outcome of a deterministic event is at least theoretically completely predictable, the outcome of a random event is not fully predictable.[4] Hence, according to this interpretation, *randomness* is tantamount to *lack of certainty*. The degree of certainty, or more generally, our *belief* about the outcome of a random event, is expressed in terms of probability. Random phenomena, however, should not be confused with chaotic phenomena, whose outcomes are deterministic but so sensitive to any imprecision in initial values to make them effectively unpredictable.

That said, the concept of probability in a philosophical sense has a subjective component, since deciding what is reasonable to believe, and what previous experi-

---

[3] As Laplace put it, "probability theory is nothing more but common sense reduced to calculation."

[4] Arguably, this definition is a little murky, similarly to how what is not conscious is labeled as being part of the subconscious.

ence or information the belief is based on, is a subjective matter. In this book, unless otherwise specified, the term probability means *subjective probability*, which is also known as *Bayesian probability*. To better understand this concept, and to be ready to fend off the customary objections against it, let us consider some simple examples.

Example 1.1 Consider the random event of tossing a coin. In general, it is quite natural to assume that the *odds* (i.e., relative probabilities) of the complementary events of getting heads or tails are equal. This is formally written as

$$P\{\text{heads}\} = P\{\text{tails}\} = \frac{1}{2},$$

where P stands for the probability of the event to occur.[5] A justification for this choice, or any other assignment of the odds, it is often argued, lies in the repeatability of the random phenomenon. After tossing the coin over and over again, N times, we may compute the relative frequency of heads,

$$f_{\text{heads}}^{(N)} = \frac{\#\text{ heads occurring}}{N}.$$

Letting N go to infinity, tacitly assuming that a limit value exists, we obtain the asymptotic value of the frequency

$$f_{\text{heads}} = \lim_{N \to \infty} f_{\text{heads}}^{(N)},$$

which can be argued to represent the probability of the event. If this number is $1/2$, the coin is said to be fair, meaning that both outcomes are equally likely to occur. The repetition experiment can be seen as an empirical test of the hypothesis that the coin is fair.

The previous example is in line with the *frequentist* definition of probability: the probability of an event is identified with its relative frequency of occurrence in an asymptotically infinite series of repeated experiments. The statistics based on this view is often referred to as *frequentist statistics*,[6] and the most common argument in support of this interpretation is its seemingly objective and empirical nature.

The interpretation of probability as a frequency works well for phenomena described in terms of repeated mutually independent experiments. This underlying assumption excludes a good part of the common use of probability. An example is the probability of an outcome of a sports event: If a given sports team has played several times against another team, the next game can hardly be viewed as an independent realization of the same random event, even though the availability of information about previous scores may help to set the odds. Assessing the

---

[5] We use P with curly braces when we describe an event in words.

[6] Frequentist statistics is also called Fisherian statistics, according to the English statistician and evolutionary biologist Ronald Fisher (1890–1962).

probabilities of different outcomes requires more information, and setting the odds for betting, for instance, is a subjective, but usually not arbitrary, process.

The following example stresses the subjective nature of probability.

Example 1.2 There is a coin about to be tossed. After a series of tests, it is agreed that the coin to be tossed can be considered fair. Consider now the following two experiments:

1. You are asked to predict the outcome of the next coin toss. In this case, nobody knows whether the result will be heads or tails.
2. You are asked to guess the outcome of a coin toss that has already occurred. In this case, the person tossing the coin checks the outcome without showing it, and therefore knows the outcome without revealing it to you.

The perceived fairness of the coin suggests that you may guess either one of the possible outcomes as you wish, the probabilities being one half for both outcomes. This consideration applies to both experiments. However, there is an important philosophical difference between the two settings.

In the first case, both subjects are equally ignorant about the outcome of the coin toss, and need to use the concept of probability to predict it. In the second case, *the tosser knows* the outcome and, unlike you, has no reason to use the concept of probability. From the tosser's point of view, you are assigning probabilities to a fully deterministic, and known, outcome. In a betting context, for instance, the tosser's advantage in the second experiment is obvious,[7] but the other person's knowledge of the outcome does not help you at all to set up the odds.

This example motivates the following important considerations about the nature of probability.

1. Whether the concept of probability is needed or not depends on the subject (you versus the coin tosser), not on the object (the coin). The probability can therefore be argued to be a property of the *subject, and not the object.*[8]
2. Probability is used when the information is perceived as incomplete. Therefore, we may define probability as an expression of *subject's lack of certainty*, or *ignorance*, arising from the lack of relevant information.
3. Whether the quantity that we consider is well-defined and deterministic, such as the face of a coin lying on the table, or a realization of a random process, such as the coin toss, the probabilistic description depends only on what we know about the quantity.

Subjective probability expresses a subject's belief about the outcome of an observation. Subjectivity is not always appreciated in science, which strives towards objectivity, in the sense of seeking to establish statements that can be shared unequivocally by every intelligent being. However, the concept of subjective probability is not in

---

[7] Of course, only a fool would be accepting to bet under those conditions!

[8] The Italian mathematician Bruno De Finetti (1906–1985) expressed this observation strongly by stating that "probability does not exist."

conflict with the scientific endeavor because subjective is not the same as arbitrary, and shared evidence tends to lead to shared beliefs. Whenever we assign probabilities to events, we should base it on the best judgement, such as previous experiments, well-established models, or solid reasoning. For instance, in the previous example, assigning equal probabilities to heads and tails was based on the presumably reliable information, and possibly on extensive experimental evidence, that the coin is fair.

We conclude the discussion with the notion of *objective chance*. The current interpretation of quantum mechanics implicitly assumes the existence of objective chance, meaning that the probabilities of events are non-judgemental. For instance, the square of the absolute value of the solution of the Schrödinger equation describing a particle in a potential field defines a probability density, predicting an objective chance of finding a particle in a given domain. However, whether the uncertainty about the value of a quantity is fundamental or ontological, as in quantum physics, or contingent or epistemological, as in the lack of a good enough measurement device, is of no consequence from the point of view of certainty: to predict an outcome, or to estimate an unknown, it is not important to know why we are ignorant.[9]

1.2 Through the Formal Theory, Lightly

Regardless of the interpretation, the theory of probability and randomness is widely accepted and presented in a standard form. The formalization of the theory is attributed to Kolmogorov.[10] Since our emphasis is on something besides theoretical considerations, we will present a light version of the fundamentals, spiced with examples and intuitive reasoning. A rigorous treatment of the matter can be found in standard textbooks on probability theory, see Notes and Comments at the end of the chapter. We begin by presenting some definitions and basic results that will be used extensively later. We will not worry too much here about the formal conditions necessary for the quantities introduced to be well defined. In particular, questions concerning, e.g., measurability or completeness will be neglected.

---

[9] The Danish physicist Niels Bohr (1885–1962), who contributed significantly to the understanding of randomness in quantum physics, wrote: "It is wrong to think that the task of physics is to find out how Nature is. Physics concerns what we say about Nature." This statement, pointing out the difference between physics and metaphysics, is in line with the subjective probability in the sense that the source of uncertainty is of no consequence. This philosophical viewpoint has been recently advanced in the theory of Quantum Bayesianism, of QBism.

[10] Andrej Nikolaevich Kolmogorov (1903–1987): a Russian mathematician, often referred to as one of the most prominent figures in the twentieth-century mathematics and the father of modern probability theory.

1.2.1 Elementary Probabilities

The standard theoretical treatment of probability starts by postulating a triplet $(\Omega, \mathscr{B}, \mathbb{P})$, called the *probability space*. Here, Ω is an abstract set called the *sample space*. The set $\mathscr{B}$ is a collection of subsets of Ω, satisfying the following conditions:

(i) $\Omega \in \mathscr{B}$,
(ii) If $A \in \mathscr{B}$, then $\Omega \setminus A \in \mathscr{B}$,
(iii) If $A_1, A_2, \ldots \in \mathscr{B}$, then $\cup_{j=1}^{\infty} A_j \in \mathscr{B}$.

A σ-algebra is a collection of subsets satisfying the conditions above. We call the set $\mathscr{B}$ the *event space*, and the individual sets in it are referred to as *events*. The third element in the probability space triplet is the *probability measure* $\mathbb{P}$, a mapping

$$\mathbb{P} : \mathscr{B} \to \mathbb{R}, \quad \mathbb{P}(E) = \text{probability of } E, \, E \subset \Omega,$$

that must satisfy the following conditions:

(i) $0 \le \mathbb{P}(E) \le 1, \quad E \in \mathscr{B}$,
(ii) $\mathbb{P}(\Omega) = 1$,
(iii) if $A_j \in \mathscr{B}$, $j = 1, 2, 3, \ldots$ with $A_j \cap A_k = \emptyset$ whenever $j \ne k$, we have

$$\mathbb{P}\left(\bigcup_{j=1}^{\infty} A_j\right) = \sum_{j=1}^{\infty} \mathbb{P}(A_j). \quad (\sigma\text{-additivity})$$

It follows from the definition that

$$\mathbb{P}(\Omega \setminus A) = 1 - \mathbb{P}(A),$$

which implies that $\mathbb{P}(\emptyset) = 0$. Moreover, if $A_1, A_2 \in \mathscr{B}$ and $A_1 \subset A_2 \subset \Omega$,

$$\mathbb{P}(A_1) \le \mathbb{P}(A_2).$$

Two events, A and B, are *independent*, if

$$\mathbb{P}(A \cap B) = \mathbb{P}(A)\mathbb{P}(B).$$

The *conditional probability* of A given B is the probability that A happens *provided* that B happens,

$$\mathbb{P}(A \mid B) = \frac{\mathbb{P}(A \cap B)}{\mathbb{P}(B)}, \quad \text{assuming that } \mathbb{P}(B) > 0.$$

The probability of A conditioned on B is never smaller than the probability of $A \cap B$, since $\mathbb{P}(B) \le 1$, and it can be much larger if $\mathbb{P}(B)$ is very small, however, never larger than 1, since

$$P(A \mid B) = \frac{\mathbb{P}(A \cap B)}{\mathbb{P}(B)} \leq \frac{\mathbb{P}(\Omega \cap B)}{\mathbb{P}(B)} = \frac{\mathbb{P}(B)}{\mathbb{P}(B)} = 1.$$

This corresponds to the intuitive idea that the conditional probability is a probability on the probability subspace where the event B has already occurred. It follows from the definition of independent events that, if A and B are mutually independent, then

$$\mathbb{P}(A \mid B) = \mathbb{P}(A), \quad \mathbb{P}(B \mid A) = \mathbb{P}(B).$$

Vice versa, if one of the above equalities holds, then by the definition of conditional probabilities A and B must be independent.

Consider two events A and B, both having a positive probability, and write the conditional probability of A given B and of B given A,

$$\mathbb{P}(A \mid B) = \frac{\mathbb{P}(A \cap B)}{\mathbb{P}(B)}, \quad \mathbb{P}(B \mid A) = \frac{\mathbb{P}(A \cap B)}{\mathbb{P}(A)}.$$

By solving for the probability $\mathbb{P}(A \cap B)$ from the former and substituting it in the latter, we obtain

$$\mathbb{P}(B \mid A) = \frac{\mathbb{P}(A \mid B)\mathbb{P}(B)}{\mathbb{P}(A)}. \tag{1.1}$$

This is *Bayes' formula* for elementary events, which is the very heart of most of the topic discussed in this book.

1.2.2 Probability Distributions and Densities

Given a sample space Ω, a real valued *random variable* X is a mapping

$$X : \Omega \to \mathbb{R},$$

which assigns to each element of Ω a real value $X(\omega)$, such that for every open set $A \subset \mathbb{R}$, $X^{-1}(A) \in \mathscr{B}$. The latter condition is expressed by saying that X is a *measurable function*. We call $x = X(\omega)$, $\omega \in \Omega$, a *realization* of X. Therefore, a random variable is a function, while the realizations of a real-valued random variable are real numbers.

For each $B \subset \mathbb{R}$, we define

$$\mu_X(B) = \mathbb{P}(X^{-1}(B)) = P\{X(\omega) \in B\},$$

and call μ_X the *probability distribution* of X, i.e., $\mu_X(B)$ is the probability of the *event* $\{\omega \in \Omega : X(\omega) \in B\}$. The probability distribution $\mu_X(B)$ measures the size of the subset of Ω mapped onto B by the random variable X.

We restrict the discussion here mostly to probability distributions that are absolutely continuous with respect to the Lebesgue measure over the reals,[11] meaning that there exists a function, the *probability density* π_X of X, such that

$$\mu_X(B) = \int_B \pi_X(x)dx. \tag{1.2}$$

A function is a probability density if it satisfies the following two conditions:

$$\pi_X(x) \geq 0, \quad \int_{\mathbb{R}} \pi_X(x)dx = 1. \tag{1.3}$$

Conversely, any function satisfying (1.3) can be viewed as a probability density of some random variable.

The *cumulative distribution function* (cdf) of a real-valued random variable is defined as

$$\Phi_X(x) = \int_{-\infty}^{x} \pi_X(x')dx' = P\{X \leq x\}.$$

The cumulative distribution function will play a very important role in performing random draws from a distribution. Observe that Φ_X is non-decreasing, and it satisfies

$$\lim_{x \to -\infty} \Phi_X(x) = 0, \quad \lim_{x \to \infty} \Phi_X(x) = 1.$$

The definition of random variables can be generalized to cover multidimensional state spaces. Given two real-valued random variables X and Y, the joint probability distribution defined over Cartesian products of sets is

$$\mu_{X,Y}(A \times B) = P\{X \in A, Y \in B\}$$
$$= \mathbb{P}(X^{-1}(A) \cap Y^{-1}(B))$$
$$= \text{the probability of the event that } X \in A$$
$$\text{and, } \textit{at the same time, } Y \in B,$$

where $A, B \subset \mathbb{R}$. Assuming that the probability distribution can be written as an integral of the form

$$\mu_{X,Y}(A \times B) = \int\int_{A \times B} \pi_{XY}(x, y)dxdy, \tag{1.4}$$

the non-negative function π_{XY} defines the *joint probability density* of the random variables X and Y. We may define a two-dimensional random variable,

---

[11] Concepts like absolute continuity of measures will play no role in these notes, and this comment added here is just to avoid unnecessary "friendly fire" from the camp of our fellow mathematicians.

$$Z = \begin{bmatrix} X \\ Y \end{bmatrix},$$

and by approximating general two-dimensional sets by unions of rectangles, we may write

$$P\{Z \in B\} = \int\int_B \pi_{XY}(x, y)dxdy = \int_B \pi_Z(z)dz, \qquad (1.5)$$

where we used the notation $\pi_{XY}(x, y) = \pi_Z(z)$, and the integral with respect to z is the two-dimensional integral, $dz = dxdy$. More generally, we define a *multivariate random variable* as a measurable mapping

$$X = \begin{bmatrix} X_1 \\ \vdots \\ X_n \end{bmatrix} : \Omega \to \mathbb{R}^n,$$

where each component X_i is an $\mathbb{R}$-valued random variable. The probability density of X is the joint probability density

$$\pi_X = \pi_{X_1 X_2 \cdots X_n} : \mathbb{R}^n \to \mathbb{R}_+$$

of its components, satisfying

$$P\{X \in B\} = \mu_X(B) = \int_B \pi_X(x)dx, \quad B \subset \mathbb{R}^n.$$

Consider two multivariate random variables $X : \Omega \to \mathbb{R}^n$ and $Y : \Omega \to \mathbb{R}^m$. Their joint probability density π_{XY} is defined in the space $\mathbb{R}^{n+m}$ similarly to (1.4), choosing first $A \subset \mathbb{R}^n$ and $B \subset \mathbb{R}^m$, and then, more generally, extending the definition to all sets in $\mathbb{R}^{n+m}$ as in (1.5). In the sequel, we will use the shorthand notation $X \in \mathbb{R}^n$ to indicate that X is an n-variate random variable, that is, its values are in $\mathbb{R}^n$.

The random variables $X \in \mathbb{R}^n$ and $Y \in \mathbb{R}^m$ are *independent* if

$$\pi_{XY}(x, y) = \pi_X(x)\pi_Y(y), \qquad (1.6)$$

in agreement with the definition of independent events. This formula gives us also a way to calculate the joint probability density of two independent random variables.

Given two not necessarily independent random variables $X \in \mathbb{R}^n$ and $Y \in \mathbb{R}^m$ with joint probability density $\pi_{XY}(x, y)$, the *marginal density* of X is the probability of X when Y may take on any value,

$$\pi_X(x) = \int_{\mathbb{R}^m} \pi_{XY}(x, y)dy.$$

In other words, the marginal density of X is simply the probability density of X without any thoughts about Y. The marginal of Y is defined analogously by the formula

$$\pi_Y(y) = \int_{\mathbb{R}^n} \pi_{XY}(x, y)dx. \tag{1.7}$$

Observe that, in general, (1.6) does not hold for marginal densities.

An important concept in Bayesian scientific computing is that of *conditioning*. Consider formula (1.7), and assume that $\pi_Y(y) \neq 0$. Dividing both sides by the scalar $\pi_Y(y)$ gives the identity

$$\int_{\mathbb{R}^n} \frac{\pi_{XY}(x, y)}{\pi_Y(y)} dx = 1.$$

Since the integrand is a non-negative function, it defines a probability density for X, for fixed y. We define the *conditional probability density* of X given Y,

$$\pi_{X|Y}(x \mid y) = \frac{\pi_{XY}(x, y)}{\pi_Y(y)}, \quad \pi_Y(y) \neq 0.$$

With some caution, and in a rather cavalier way, one can interpret $\pi_{X|Y}$ as the probability density of X, assuming that the random variable Y takes on the value $Y = y$. The conditional density of Y given X is defined similarly as

$$\pi_{Y|X}(y \mid x) = \frac{\pi_{XY}(x, y)}{\pi_X(x)}, \quad \pi_X(x) \neq 0.$$

Observe that the symmetric roles of X and Y imply that

$$\pi_{XY}(x, y) = \pi_{X|Y}(x \mid y)\pi_Y(y) = \pi_{Y|X}(y \mid x)\pi_X(x), \tag{1.8}$$

leading to the important identity known as *Bayes' formula* for probability densities,

$$\pi_{X|Y}(x \mid y) = \frac{\pi_{Y|X}(y \mid x)\pi_X(x)}{\pi_Y(y)}. \tag{1.9}$$

This identity will play a central role in the rest of the book.

Graphical interpretation of the marginal and conditional densities are presented in Fig. 1.1.

1.2.3 Expectation and Covariance

Given a random variable $X \in \mathbb{R}^n$ with probability density π_X, and a function $f : \mathbb{R}^n \to \mathbb{R}$, we define the *expectation* of $f(X)$ as

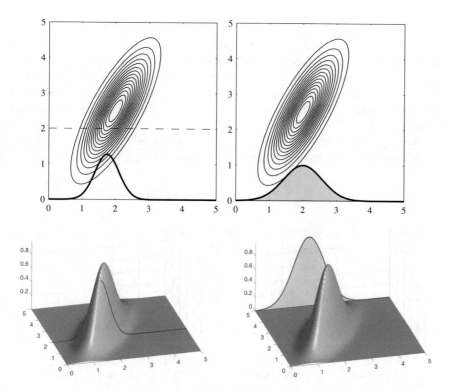

Fig. 1.1 Conditional density (left) and marginal density (right) visualized for two real-valued random variables X and Y. The ellipsoids in the top panels are equiprobability curves of the joint probability density $\pi_{XY}(x, y)$. The lower panels show 3D renditions of the conditional (left) and marginal densities (right)

$$\mathrm{E}\big(f(X)\big) = \int_{\mathbb{R}^n} f(x)\pi_X(x)dx,$$

assuming that the integral converges. Intuitively, the expectation can be thought of as the value that one could expect the quantity $f(X)$ to take based on the knowledge of the probability density. In the following, we consider some important and useful expectations.

Given a random variable $X \in \mathbb{R}$, its expected value, or mean value, is the center of mass of the probability distribution defined as

$$\mathrm{E}\big(X\big) = \overline{x} = \int_{\mathbb{R}} x\, \pi_X(x)dx \in \mathbb{R},$$

assuming that the integral is convergent. The *variance* of the random variable X is the expectation of the squared deviation from the expectation,

$$\text{var}(X) = \mathrm{E}\big((X - \overline{x})^2\big) = \sigma_X^2 = \int_{\mathbb{R}} (x - \overline{x})^2 \, \pi_X(x) dx, \qquad (1.10)$$

provided that the integral is finite. It is the role of the variance to measure how distant from the expectation the values taken on by the random variable are. A small variance means that the random variable takes on mostly values close to the expectation, while a large variance means the opposite. The square root σ_X of the variance is the *standard deviation* of X.

The expectation and the variance are also known as the first two moments of a probability density function. The kth moment is defined as

$$\mathrm{E}\big((X - \overline{x})^k\big) = \int_{\mathbb{R}} (x - \overline{x})^k \, \pi_X(x) dx,$$

assuming that the integral converges. The third moment ($k = 3$) is related to the *skewness* of the probability density, defined as

$$\text{skew}(X) = \frac{\mathrm{E}\big((X - \overline{x})^3\big)}{\sigma_X^3},$$

and the fourth ($k = 4$) is related to the *kurtosis*,

$$\text{kurt}(X) = \frac{\mathrm{E}\big((X - \overline{x})^4\big)}{\sigma_X^4}.$$

The mean and the variance lend themselves to an immediate generalization to multivariate random variables. Given $X \in \mathbb{R}^n$, the mean of X is the vector in $\mathbb{R}^n$,

$$\overline{x} = \int_{\mathbb{R}^n} x \, \pi_X(x) dx = \begin{bmatrix} \overline{x}_1 \\ \vdots \\ \overline{x}_n \end{bmatrix} \in \mathbb{R}^n,$$

or, component-wise,

$$\overline{x}_j = \int_{\mathbb{R}^n} x_j \, \pi_X(x) dx \in \mathbb{R}, \quad 1 \le j \le n.$$

We may calculate the variance of each component in a straightforward manner, by defining

$$\text{var}(X_j) = \int_{\mathbb{R}^n} (x_j - \overline{x}_j)^2 \, \pi_X(x) dx, \quad 1 \le j \le n. \qquad (1.11)$$

A more versatile quantity that encompasses the above definition while also accounting for the higher dimensionality of the random variable is the *covariance matrix*. By definition, the covariance is an $n \times n$ matrix with elements

$$\text{cov}(X)_{ij} = \int_{\mathbb{R}^n} (x_i - \bar{x}_i)(x_j - \bar{x}_j)\,\pi_X(x)dx \in \mathbb{R}, \quad 1 \le i, j \le n.$$

Alternatively, we can define the covariance using vector notation as

$$\text{cov}(X) = \int_{\mathbb{R}^n} (x - \bar{x})(x - \bar{x})^{\mathsf{T}}\,\pi_X(x)dx \in \mathbb{R}^{n \times n}.$$

Observe that the variances of the components given by (1.11) are the diagonal entries of the covariance matrix.

The expectation of functions depending on only one component of a multivariate random variable can be calculated by using the corresponding marginal density. Consider a function $f(X_i)$ of the ith component of X. Denoting by $x_i' \in \mathbb{R}^{n-1}$ the vector obtained by deleting the ith component from x,

$$\begin{aligned}
\mathrm{E}\big(f(X_i)\big) &= \int_{\mathbb{R}^n} f(x_i)\,\pi_X(x)dx \\
&= \int_{\mathbb{R}} f(x_i) \underbrace{\left(\int_{\mathbb{R}^{n-1}} \pi_X(x_i, x_i')dx_i' \right)}_{=\pi_{X_i}(x_i)} dx_i \\
&= \int_{\mathbb{R}} f(x_i)\,\pi_{X_i}(x_i)dx_i.
\end{aligned}$$

An immediate consequence of this observation is that, by letting $f(x_i) = x_i$, we may write the mean and the diagonal entries of the covariance matrix as

$$\bar{x}_i = \int_{\mathbb{R}} x_i\,\pi_{X_i}(x_i)dx_i,$$

$$\text{var}(X_i) = \text{cov}(X)_{ii} = \int_{\mathbb{R}} (x_i - \bar{x}_i)^2\,\pi_{X_i}(x_i)dx_i.$$

The properties of the covariance matrix will be discussed in more detail in the next chapter after the introduction of some tools from linear algebra.

The definition of conditional density leads naturally to the concept of *conditional expectation* or *conditional mean*. Given two random variables $X \in \mathbb{R}^n$ and $Y \in \mathbb{R}^m$, we define

$$\mathrm{E}(X \mid y) = \int_{\mathbb{R}^n} x\,\pi_{X|Y}(x \mid y)dx.$$

To compute the expectation of X via its conditional expectation, observe that

$$\mathrm{E}(X) = \int_{\mathbb{R}^n} x\,\pi_X(x)dx = \int_{\mathbb{R}^n} x \left(\int_{\mathbb{R}^m} \pi_{XY}(x, y)dy \right) dx,$$

and, substituting (1.8) into this expression, we obtain

$$E(X) = \int_{\mathbb{R}^n} x \left(\int_{\mathbb{R}^m} \pi_{X|Y}(x \mid y)\pi_Y(y)dy \right) dx \tag{1.12}$$

$$= \int_{\mathbb{R}^m} \left(\int_{\mathbb{R}^n} x\pi_{X|Y}(x \mid y)dx \right) \pi_Y(y)dy = \int_{\mathbb{R}^m} E(X \mid y)\pi_Y(y)dy.$$

We conclude this section with some examples.

Example 1.3 The expectation of a random variable, in spite of its name, is not always a value that we can expect a realization takes on. Whether this is the case or not, it depends on the distribution of the random variable. To clarify what we mean, consider two real-valued random variables X_1 and X_2 with probability densities

$$\pi_{X_1}(x) = \begin{cases} 1, & |x| < 1/2, \\ 0, & |x| \geq 1/2, \end{cases} \tag{1.13}$$

and

$$\pi_{X_2}(x) = \begin{cases} 1, & |x - 1| < 1/4, \\ 1, & |x + 1| < 1/4, \\ 0, & \text{elsewhere.} \end{cases}$$

It is straightforward to check that

$$E(X_1) = E(X_2) = 0,$$

and that, while the expected value is a possible value for X_1, this is not the case for any realization of X_2. This simple consideration serves as a warning against the danger of describing a random variable only through its expectation.

Example 1.4 Random variables waiting for the train: Assume that every day, except on Sundays, a train for your destination leaves every S minutes from the station. On Sundays, the interval between trains is $2S$ minutes. You arrive at the station with no information about the trains' timetable. What is your expected waiting time?

To address the problem, we define two integer-valued random variables: W is the time variable measured in units of days, $0 \leq W \leq 7$, so that, for instance, the event $\{0 \leq W < 1\}$ is Monday. The second random variable T indicates the waiting time. The conditional distribution of T given W is then

$$\pi_{T|W}(t \mid w) = \frac{1}{S}\chi_S(t), \quad \chi_S(t) = \begin{cases} 1, & 0 \leq t < S, \\ 0 & \text{otherwise} \end{cases} \quad \text{if } 0 \leq w < 6, \tag{1.14}$$

and

$$\pi_{T|W}(t \mid w) = \frac{1}{2S}\chi_{2S}(t), \quad \text{if } 6 \leq w < 7. \tag{1.15}$$

If you are absolutely sure that it is not Sunday, you can express your belief about the waiting time using the conditional density (1.14), leading to the expected waiting time

$$E(T \mid w) = \int_{\mathbb{R}} t \, \pi_{T|W}(t \mid w) \, dt = \frac{1}{S} \int_0^S t \, dt = \frac{S}{2}, \quad 0 \le w < 6.$$

The expected waiting time with certainty that it is Sunday is

$$E(T \mid w) = \int_{\mathbb{R}} t \, \pi_{T|W}(t \mid w) \, dt = \frac{1}{2S} \int_0^{2S} t \, dt = S, \quad 6 \le w < 7.$$

If you have no idea which day of the week it is,[12] you may give equal probability to each weekday, defining the probability density of W as

$$\pi_W(w) = \frac{1}{7}, \quad 0 \le w \le 7.$$

The expectation of T can be computed according to (1.12), yielding

$$
\begin{aligned}
E(T) &= \int_{\mathbb{R}} E(T \mid w) \pi_W(w) \, dw \\
&= \frac{1}{7} \int_0^6 E(T \mid w) \, dw + \frac{1}{7} \int_6^7 E(T \mid w) \, dw \\
&= \frac{1}{7} \int_0^6 \frac{S}{2} \, dw + \frac{1}{7} \int_6^7 S \, dw = \frac{4}{7} S.
\end{aligned}
$$

1.2.4 Change of Variables in Probability Densities

It is not uncommon that we are interested in the probability density of a random variable that is a function of another random variable with known probability density. Therefore, we need to perform a change of variables, keeping in mind that we are working with probability densities. We start by solving the problem in the one-dimensional case, then consider the general multivariate case.

Assume that we have two real-valued random variables X, Z that are related to each other through a functional relation

$$X = \phi(Z),$$

---

[12] The probability of this to happen seems to be much larger for mathematicians than for other individuals.

where $\phi : \mathbb{R} \to \mathbb{R}$ is a one-to-one mapping. For simplicity, assume that ϕ is strictly increasing and differentiable, so that $\phi'(z) > 0$. If the probability density function π_X of X is given, what is the corresponding density π_Z of Z?

First, note that since ϕ is increasing, for any values $a < b$, we have

$$a < Z < b \quad \text{if and only if} \quad a' = \phi(a) < \phi(Z) = X < \phi(b) = b',$$

therefore

$$P\{a' < X < b'\} = P\{a < Z < b\}.$$

Equivalently, the probability density of Z satisfies

$$\int_a^b \pi_Z(z)dz = \int_{a'}^{b'} \pi_X(x)dx.$$

Performing a change of variables in the integral on the right,

$$x = \phi(z), \quad dx = \frac{d\phi}{dz}(z)dz,$$

we obtain

$$\int_a^b \pi_Z(z)dz = \int_a^b \pi_X(\phi(z))\frac{d\phi}{dz}(z)dz.$$

This holds for all a and b, and therefore we arrive at the conclusion that

$$\pi_Z(z) = \pi_X(\phi(z))\frac{d\phi}{dz}(z).$$

In the derivation above, we assumed that ϕ was increasing. If it is decreasing, the derivative is negative. In general, since the density needs to be non-negative, we write

$$\pi_Z(z) = \pi_X(\phi(z))\left|\frac{d\phi}{dz}(z)\right|.$$

The above reasoning for one-dimensional random variables can be extended to multivariate random variables as follows. Let $X \in \mathbb{R}^n$ and $Z \in \mathbb{R}^n$ be two random variables such that

$$X = \phi(Z),$$

where $\phi : \mathbb{R}^n \to \mathbb{R}^n$ is a one-to-one differentiable mapping. Consider a set $B \subset \mathbb{R}^n$, and let $B' = \phi(B) \subset \mathbb{R}^n$ be its image in the mapping ϕ. Then we may write

$$\int_B \pi_Z(z)dz = \int_{B'} \pi_X(x)dx.$$

We perform the change of variables $x = \phi(z)$ in the latter integral, remembering that

$$dx = \left| \det\big(D\phi(z)\big) \right| dz,$$

where $D\phi(x)$ is the Jacobian of the mapping ϕ,

$$D\phi(z) = \begin{bmatrix} \frac{\partial \phi_1}{\partial z_1} & \cdots & \frac{\partial \phi_1}{\partial z_n} \\ \vdots & & \vdots \\ \frac{\partial \phi_n}{\partial z_1} & \cdots & \frac{\partial \phi_n}{\partial z_n} \end{bmatrix} \in \mathbb{R}^{n \times n},$$

and its determinant, the Jacobian determinant, expresses the local volume scaling of the mapping ϕ. Occasionally, the Jacobian determinant is written in a suggestive form to make it formally similar to the one-dimensional equivalent,

$$\frac{\partial \phi}{\partial z} = \det\big(D\phi(z)\big).$$

With this notation,

$$\int_B \pi_Z(z)dz = \int_{B'} \pi_X(x)dx = \int_B \pi_X(\phi(z)) \left| \frac{\partial \phi}{\partial z} \right| dz$$

for all $B \subset \mathbb{R}^n$, and we arrive at the conclusion that

$$\pi_Z(z) = \pi_X\big(\phi(z)\big) \left| \frac{\partial \phi}{\partial z} \right|.$$

This is the change of variables formula for probability densities.

Notes and Comments

The reference to reverend Thomas Bayes' classical essay is [3]. The Welsh philosopher and mathematician Richard Price had an important role in editing and discussing Bayes' work. The motivations of Thomas Bayes to get engaged in the kind of mathematics and statistics that led him to the concept of inverse probability are not fully known, some authors believing that Bayes was interested in refuting arguments for not believing in miracles advocated by philosophers like David Hume. An English translation of Laplace's Memoir mentioned in this chapter can be found in the reference [70].

For a groundbreaking article promoting the Bayesian methods in modern inverse problems, see [75]. The books [18, 49, 74] develop the theory of computational inverse problems further, discussing several of the themes of this book.

An inspirational reference for the topic of subjective probability is [46]. The citation of de Finetti claiming that probability does not exist can be found in his own book on probability [26]. We also encourage to read some of the discussion concerning objectivity and Bayesian philosophy, see, e.g., [4] and the related commentaries.

A good standard reference for probability theory is [5]. For a more elementary but thorough introduction to probability theory, we refer to [1]. An extensive discussion concerning the intricacies of defining conditional distributions can be found in the monograph [36].

Chapter 2
Linear Algebra

There is no question that linear algebra is one of the cornerstones of scientific computing. Moreover, linear algebra provides an effective language for dealing with multidimensional phenomena, including multivariate statistics that without this language would become awkward and cumbersome. Instead of collecting all the linear algebra definitions and results that will be needed in a comprehensive primer, we introduce them gradually throughout the exposition, so as not to overwhelm the reader and to present examples of how these ideas are utilized. That said, for the benefits of readers with less traditional mathematical and statistical background, in this chapter, we collect some of the most foundational linear algebra definitions and results that will be used extensively in the remaining chapters.

2.1 Vectors and Matrices

In this book, we restrict the discussion of linear algebra to real vectors and matrices. Given x in $\mathbb{R}^n$ with Cartesian coordinates $(x_1, x_2, \ldots, x_n)$, we identify the point with its position vector, interpreted as a column vector

$$x = \begin{bmatrix} x_1 \\ \vdots \\ x_n \end{bmatrix}.$$

The corresponding "row vector" is obtained by ordering the coordinates in a row, an operation that is a special instance of the transposition to be formally defined later,

$$x^\mathsf{T} = \begin{bmatrix} x_1 & \cdots & x_n \end{bmatrix}.$$

© The Author(s), under exclusive license to Springer Nature Switzerland AG 2023
D. Calvetti and E. Somersalo, *Bayesian Scientific Computing*, Applied Mathematical
Sciences 215, https://doi.org/10.1007/978-3-031-23824-6_2

The addition of two vectors of same size[1] is defined component-wise, as is the operation of scalar multiplication, where all components of a vector are multiplied by the same scalar. A linear combination of a set of vectors is the sum of scalar multiples of the vectors.

The inner products of two vectors $x, y \in \mathbb{R}^n$ is the scalar

$$\sum_{j=1}^{n} x_j y_j = x^\mathsf{T} y = y^\mathsf{T} x, \quad \text{where} \quad x = \begin{bmatrix} x_1 \\ \vdots \\ x_n \end{bmatrix}, \quad y = \begin{bmatrix} y_1 \\ \vdots \\ y_n \end{bmatrix}.$$

This definition of inner product is related to the Euclidian length, or ℓ^2-norm of the vector defined through its square as

$$\|x\|^2 = \sum_{j=1}^{n} x_j^2 = x^\mathsf{T} x.$$

Two vectors $x, y \in \mathbb{R}^n$ are *orthogonal* with respect to the inner product defined above if and only if

$$x^\mathsf{T} y = 0,$$

a definition that generalizes the notion of orthogonality, or perpendicularity of vectors in $\mathbb{R}^2$ and $\mathbb{R}^3$.

Matrices are two-dimensional arrays of real numbers organized in rows and columns,

$$A = \begin{bmatrix} a_{11} & \cdots & a_{1n} \\ \vdots & & \vdots \\ a_{m1} & \cdots & a_{mn} \end{bmatrix} \in \mathbb{R}^{m \times n}. \tag{2.1}$$

The matrix–vector product of a matrix $A \in \mathbb{R}^{m \times n}$ and a vector $x \in \mathbb{R}^n$ is the vector

$$Ax \in \mathbb{R}^m$$

with entries

$$(Ax)_k = \sum_{j=1}^{n} a_{kj} x_j, \quad 1 \le k \le m.$$

By regarding the matrix A as the collection of its column vectors,

$$A = \begin{bmatrix} | & | & & | \\ a_1 & a_2 & \cdots & a_n \\ | & | & & | \end{bmatrix}, \quad a_j \in \mathbb{R}^m,$$

---

[1] In this case, the term *size* refers to the number of components and to their ordering.

the matrix–vector product can be viewed as the linear combination of the columns of the matrix, with the entries of the vector as coefficients,

$$A x = \sum_{j=1}^{n} x_j a_j. \tag{2.2}$$

The column-wise interpretation of a matrix allows us to generalize the matrix–vector product to a matrix–matrix product. Given $A \in \mathbb{R}^{m \times n}$, $B \in \mathbb{R}^{n \times k}$, with

$$B = \begin{bmatrix} b_1 & b_2 & \cdots & b_k \end{bmatrix}, \quad b_j \in \mathbb{R}^n,$$

the product of the two matrices A and B is

$$C = AB = \begin{bmatrix} A b_1 & A b_2 & \cdots & A b_k \end{bmatrix} \in \mathbb{R}^{m \times k}.$$

It can be verified in a straightforward way that this column-centric definition of matrix–matrix product coincides with the traditional component-centric definition in terms of the entries of the two matrices,

$$C = AB \in \mathbb{R}^{m \times k}, \quad c_{ij} = \sum_{\ell=1}^{n} a_{i\ell} b_{\ell j}, \quad 1 \le i \le m, \quad 1 \le j \le k.$$

We can use this definition, together with the interpretation of vectors as special matrices, to define the *outer product* of two vectors $x \in \mathbb{R}^m$, $y \in \mathbb{R}^n$, not necessarily of the same size, as

$$x y^{\mathsf{T}} = \begin{bmatrix} x_1 \\ \vdots \\ x_m \end{bmatrix} \begin{bmatrix} y_1 & \cdots & y_n \end{bmatrix} = \begin{bmatrix} x_1 y_1 & \cdots & x_1 y_n \\ \vdots & & \vdots \\ x_m y_1 & \cdots & x_m y_n \end{bmatrix} \in \mathbb{R}^{m \times n}.$$

The *transposition* of a matrix is the operation that swaps the rows and columns of the matrix. For example, if $A \in \mathbb{R}^{m \times n}$ is as in (2.1), the transpose of A is

$$A^{\mathsf{T}} = \begin{bmatrix} a_{11} & \cdots & a_{m1} \\ \vdots & & \vdots \\ a_{1n} & \cdots & a_{mn} \end{bmatrix} \in \mathbb{R}^{n \times m}.$$

An important property of the transpose which is straightforward to verify is that it is distributive over the product, provided that we reverse the order of the factors,

$$(AB)^{\mathsf{T}} = B^{\mathsf{T}} A^{\mathsf{T}}.$$

A central concept in linear algebra is that of linear independence. The k vectors $a_1, a_2, \ldots, a_k \in \mathbb{R}^n$ are *linearly independent* if and only if the only way to get the zero vector as their linear combination is if all coefficients are equal to zero,

$$c_1 a_1 + c_2 a_2 + \ldots + c_k a_k = 0 \Leftrightarrow c_1 = c_2 = \ldots = c_k = 0. \tag{2.3}$$

If a set of vectors are linearly independent, it is not possible to express any one of them as a linear combination of the others. The linear independence of a set of vectors can be expressed concisely in matrix notation. Letting

$$\mathsf{A} = \begin{bmatrix} a_1 \ a_2 \ \cdots \ a_k \end{bmatrix} \in \mathbb{R}^{n \times k}, \quad c = \begin{bmatrix} c_1 \\ \vdots \\ c_k \end{bmatrix},$$

condition (2.3) can be formulated as

$$\mathsf{A}c = 0_n \Leftrightarrow c = 0_k,$$

where 0_n (0_k) is the vector in $\mathbb{R}^n$ $(\mathbb{R}^k)$ with all zero entries. In the sequel, the dimension of a null vector is usually not indicated.

Given k vectors $a_1, \ldots a_k$, the subspace spanned by these vectors is defined as the set of all their linear combinations,

$$H = \text{span}\{a_1, \ldots, a_k\} = \{x \in \mathbb{R}^n \mid x = c_1 a_1 + \ldots c_k a_k$$
$$\text{for some } c_1, \ldots, c_k \in \mathbb{R}\},$$

and the vectors $a_1, \ldots, a_k$ are called a spanning set of H. If the vectors $a_1, \ldots, a_k$ are a spanning set of H and they are linearly independent, they form a *basis* of H. If a vector space has a basis with a finite number of elements, every basis has the same number of elements. The *dimension* of a subspace $H \subset \mathbb{R}^n$ is the maximum number of independent vectors that span the subspace H, or equivalently, the number of elements in a basis.

The *identity matrix* of size n is an $n \times n$ matrix with zero off-diagonal entries and ones in the diagonal,

$$\mathsf{I}_n = \begin{bmatrix} 1 & & \\ & \ddots & \\ & & 1 \end{bmatrix}.$$

A matrix $\mathsf{A} \in \mathbb{R}^{n \times n}$ is invertible if there exists a matrix $\mathsf{A}^{-1} \in \mathbb{R}^{n \times n}$, called the *inverse* of A, such that

$$\mathsf{A}^{-1}\mathsf{A} = \mathsf{A}\mathsf{A}^{-1} = \mathsf{I}_n = \begin{bmatrix} 1 & & \\ & \ddots & \\ & & 1 \end{bmatrix},$$

If the columns of an $n \times n$ matrix U are mutually orthogonal vectors and have Euclidean length one, then U is invertible and $\mathsf{U}^{-1} = \mathsf{U}^\mathsf{T}$. In this case, we say that U is an *orthogonal matrix*. The orthogonality of the matrix is often expressed through the equations

$$\mathsf{U}^\mathsf{T}\mathsf{U} = \mathsf{U}\mathsf{U}^\mathsf{T} = \mathsf{I}_n.$$

2.1.1 The Singular Value Decomposition

The *singular value decomposition* (SVD) is among the most powerful and versatile matrix factorizations, and the method of choice to understand the properties of matrices, whether considered on their own, or as operators that multiply vectors. Every matrix $\mathsf{A} \in \mathbb{R}^{m \times n}$ can be expressed as the product of three matrices,

$$\mathsf{A} = \mathsf{U}\mathsf{D}\mathsf{V}^\mathsf{T},$$

where $\mathsf{U} \in \mathbb{R}^{m \times m}$ and $\mathsf{V} \in \mathbb{R}^{n \times n}$ are orthogonal matrices, that is, $\mathsf{U}^{-1} = \mathsf{U}^\mathsf{T}$, $\mathsf{V}^{-1} = \mathsf{V}^\mathsf{T}$, and $\mathsf{D} \in \mathbb{R}^{m \times n}$ is a diagonal matrix with nonnegative entries. In particular, when $n \neq m$,

$$\mathsf{D} = \begin{bmatrix} d_1 & & & \\ & d_2 & & \\ & & \ddots & \\ & & & d_n \\ & \mathsf{O}_{(m-n) \times n} & \end{bmatrix}, \text{ if } m > n,$$

and

$$\mathsf{D} = \begin{bmatrix} d_1 & & & & \\ & d_2 & & & \mathsf{O}_{m \times (n-m)} \\ & & \ddots & & \\ & & & d_m & \end{bmatrix}, \text{ if } m < n,$$

where $\mathsf{O}_{k \times \ell}$ is a null matrix of the indicated size. We write compactly,

$$\mathsf{D} = \mathrm{diag}(d_1, d_2, \ldots, d_{\min(m,n)}), \quad d_j \geq 0,$$

when the dimensions of the matrix are understood by the context. By convention, the diagonal entries $d_j \geq 0$, called the singular values of A, appear in decreasing order. The ratio of the largest to the smallest positive singular value is the *condition number* of the matrix.

The power of the SVD as an analytical tool comes from the fact that it makes it possible to not only diagonalize every matrix, but to ensure that the diagonal entries are all real and nonnegative, thus amenable to concrete interpretations. Moreover, because of the orthogonality of the matrices U and V, many of the properties of the rows and columns of A, regarded as vectors in the appropriate spaces, can be deduced from the matrix D.

In the next subsection, we define four important spaces of vectors associated with a matrix A, and we establish a relation between their dimensions and the singular values of the matrix.

2.1.2 The Four Fundamental Subspaces

The analysis of the mapping properties of a matrix can be understood geometrically by introducing certain associated subspaces.

Given a matrix $A \in \mathbb{R}^{m \times n}$, we introduce the following two important subspaces:

(i)   The *null space* of A, denoted by $\mathcal{N}(A)$, defined as

$$\mathcal{N}(A) = \{x \in \mathbb{R}^n \mid Ax = 0\}.$$

(ii)   The *range* of A, denoted by $\mathcal{R}(A)$, defined as

$$\mathcal{R}(A) = \{y \in \mathbb{R}^m \mid y = Ax \text{ for some } x \in \mathbb{R}^n\}.$$

The interpretation (2.2) of matrix–vector products provides another useful characterization of $\mathcal{R}(A)$. If $a_1, \ldots, a_n$ are the columns of A, then

$$\mathcal{R}(A) = \text{span}\{a_1, \ldots, a_n\} = \text{ the set of all}$$
$$\text{linear combinations of columns of } A.$$

It can be shown that if the matrix A has r nonzero singular values, then

$$\mathcal{R}(A) = \text{span}\{u_1, \ldots, u_r\},$$

where $u_1, \ldots, u_r$ are the first r columns of U, and

$$\mathcal{N}(A) = \text{span}\{v_{r+1}, \ldots, v_n\},$$

where $v_{r+1}, \ldots, v_n$ are the last $n - r$ columns of V.

The importance of these two subspaces becomes clear when we consider the solvability of a linear system of the form

$$Ax = b, \quad A \in \mathbb{R}^{m \times n}. \tag{2.4}$$

It follows from the definitions of null space and range that:

1. If $\mathcal{N}(\mathbf{A}) \neq \{0\}$, the solution of (2.4), if it exists, is non-unique: If $x^* \in \mathbb{R}^n$ is a solution and $0 \neq x_0 \in \mathcal{N}(\mathbf{A})$, then

$$\mathbf{A}(x^* + x_0) = \mathbf{A}x^* + \mathbf{A}x_0 = \mathbf{A}x^* = b,$$

   showing that $x^* + x_0$ is also a solution.
2. If $\mathcal{R}(\mathbf{A}) \neq \mathbb{R}^m$, the linear equation (2.4) does not always have a solution. In fact, if $b \notin \mathcal{R}(\mathbf{A})$, then, by definition, there is no $x \in \mathbb{R}^n$ such that (2.4) holds.

We are now ready to introduce two more subspaces associated with $\mathbf{A}$:

(iii) The *null space* of $\mathbf{A}^\mathsf{T}$, denoted by $\mathcal{N}(\mathbf{A}^\mathsf{T})$, defined as

$$\mathcal{N}(\mathbf{A}^\mathsf{T}) = \{x \in \mathbb{R}^m \mid \mathbf{A}^\mathsf{T}x = 0\}.$$

(iv) The *range* of $\mathbf{A}^\mathsf{T}$, denoted by $\mathcal{R}(\mathbf{A}^\mathsf{T})$, defined as

$$\mathcal{R}(\mathbf{A}^\mathsf{T}) = \{y \in \mathbb{R}^n \mid y = \mathbf{A}^\mathsf{T}x \text{ for some } x \in \mathbb{R}^m\}.$$

It can be shown that
$$\mathcal{R}(\mathbf{A}^\mathsf{T}) = \text{span}\{v_1, \ldots, v_r\},$$

where $v_1, \ldots, v_r$ are the first r columns of $\mathbf{V}$, and

$$\mathcal{N}(\mathbf{A}^\mathsf{T}) = \text{span}\{u_{r+1}, \ldots, u_m\},$$

where $u_{r+1}, \ldots, u_m$ are the last $m - r$ columns of $\mathbf{U}$.

To illustrate the significance of these two subspaces, let us take a closer look at the null space. If $x \in \mathcal{N}(\mathbf{A})$, then, by definition, $\mathbf{A}x = 0$. Moreover, for any $y \in \mathbb{R}^m$, we have
$$0 = y^\mathsf{T}\mathbf{A}x = (y^\mathsf{T}\mathbf{A})x = (\mathbf{A}^\mathsf{T}y)^\mathsf{T}x \text{ for all } y \in \mathbb{R}^m.$$

Observing that vectors of the form $\mathbf{A}^\mathsf{T}y$ represent all possible vectors in the subspace $\mathcal{R}(\mathbf{A}^\mathsf{T}) \subset \mathbb{R}^n$, we conclude that every vector in the null space $\mathcal{N}(\mathbf{A})$ must be orthogonal to every vector in $\mathcal{R}(\mathbf{A}^\mathsf{T})$. It is customary to write

$$\mathcal{N}(\mathbf{A}) \perp \mathcal{R}(\mathbf{A}^\mathsf{T}).$$

Likewise, by interchanging the roles of $\mathbf{A}$ and $\mathbf{A}^\mathsf{T}$ above, we arrive at the conclusion that
$$\mathcal{N}(\mathbf{A}^\mathsf{T}) \perp \mathcal{R}(\mathbf{A}).$$

In fact, it can be shown that every vector orthogonal to $\mathcal{R}(\mathbf{A}^\mathsf{T})$ is in $\mathcal{N}(\mathbf{A})$, which is expressed by writing

$$\begin{aligned}
\mathcal{N}(\mathbf{A}) &= \mathcal{R}(\mathbf{A}^\mathsf{T})^\perp \\
&= \{x \in \mathbb{R}^n \mid x \perp z \text{ for every } z \in \mathcal{R}(\mathbf{A}^\mathsf{T})\} \\
&= \text{orthocomplement of } \mathcal{R}(\mathbf{A}^\mathsf{T}).
\end{aligned}$$

The subspaces $\mathcal{R}(\mathbf{A}), \mathcal{N}(\mathbf{A}), \mathcal{R}(\mathbf{A}^\mathsf{T}), \mathcal{N}(\mathbf{A}^\mathsf{T})$ are commonly referred to as the *four fundamental subspaces* of $\mathbf{A}$. It is possible to decompose $\mathbb{R}^n$ in terms of two orthogonal subspaces, so that every $x \in \mathbb{R}^n$ admits a unique representation

$$x = u + v, \quad u \in \mathcal{N}(\mathbf{A}), \quad v \in \mathcal{R}(\mathbf{A}^\mathsf{T}).$$

In this case, we say that $\mathbb{R}^n$ is the *direct sum* of $\mathcal{N}(\mathbf{A})$ and $\mathcal{R}(\mathbf{A}^\mathsf{T})$, and we write

$$\mathbb{R}^n = \mathcal{N}(\mathbf{A}) \oplus \mathcal{R}(\mathbf{A}^\mathsf{T}). \tag{2.5}$$

Not surprisingly, an analogous result holds for $\mathbb{R}^m$,

$$\mathbb{R}^m = \mathcal{N}(\mathbf{A}^\mathsf{T}) \oplus \mathcal{R}(\mathbf{A}). \tag{2.6}$$

The *rank* of a matrix is the maximum number of linearly independent columns of $\mathbf{A}$ or, equivalently, the dimension of the range of $\mathbf{A}$, equaling the number of nonzero singular values of $\mathbf{A}$.

While not immediately obvious from the definition of the four fundamental subspaces, the connection with the SVD can be used to show that the maximum number of independent columns in $\mathbf{A}$ and $\mathbf{A}^\mathsf{T}$ is the same, therefore

$$\mathrm{rank}(\mathbf{A}^\mathsf{T}) = \dim(\mathcal{R}(\mathbf{A}^\mathsf{T})) = \dim(\mathcal{R}(\mathbf{A})) = \mathrm{rank}(\mathbf{A}).$$

This is often stated by saying that the column rank and the row rank of a matrix must coincide. This important result, together with (2.5) and (2.6), implies that in order for the dimensions to match, we must have

$$\dim(\mathcal{N}(\mathbf{A})) = n - \mathrm{rank}(\mathbf{A}), \quad \dim(\mathcal{N}(\mathbf{A}^\mathsf{T})) = m - \mathrm{rank}(\mathbf{A}^\mathsf{T}).$$

These results will be very handy when analyzing certain linear systems arising in the computations.

2.2 Solving Linear Systems

A core operation in scientific computing is the numerical solution of linear systems of the form (2.4). There are two basically different approaches, direct methods that depend on matrix factorizations, and iterative solvers that generate a sequence of approximations converging towards the solution. The concept of a solution of a linear system is not immediately clear; therefore, we begin by clarifying the meaning of a solution.

2.2.1 What Is a Solution?

If the matrix A in (2.4) is square and invertible, then for any $b \in \mathbb{R}^m$, the solution exists, is unique and we can write it formally as

$$x = A^{-1}b. \tag{2.7}$$

As soon as the matrix A fails to be invertible, the question of what is meant with *solution* becomes very pertinent, and the answer less categorical and obvious.

In general, if the linear system (2.4) is not square, formula (2.7) is meaningless, and the word "solution" may have a different meaning depending on the values of m and n, and on the properties of A and b. More specifically,

1. if $n > m$, the problem is *underdetermined*, i.e., there are more unknowns than linear equations to determine them;
2. if $n = m$, the problem is *formally determined*, i.e., the number of unknowns equals the number of equations;
3. if $n < m$, the problem is *overdetermined*, i.e., the number of equations exceeds that of the unknowns.

The meaning of the word "solution" differs in each of these cases, and what that implies is illustrated graphically in Fig. 2.1.

If $m < n$, the number of unknowns exceeds the number of equations. Since, in general, we can expect to be able to determine at most as many unknowns as linear equations, this means that at least $n - m$ components will not be determined without additional information about the solution. One way to overcome the problems

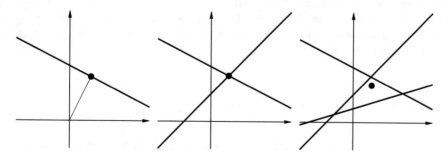

Fig. 2.1 Meaning of a solution. Left: minimum norm solution of an underdetermined system ($n = 2, m = 1$). Middle: exact solution of a non-singular system ($n = m = 2$). Right: least squares solution of an overdetermined system ($n = 2, m = 3$)

is to assign the values of $n - m$ of the unknowns, thus reducing the number of the unknowns so as to match the number of linear equations, in which case the problem is reduced to the solution of a formally determined linear system. This approach will work best if there is a reasonable criterion to selected the values of the unknowns. Another popular approach, that produces what is known as the *minimum norm solution*, is to select, among all possible solutions, the one that is closest to the origin in the Euclidean sense, i.e.,

$$x_{MN} = \arg\min \left\{ \|x\| \mid Ax = b \right\}.$$

If $m > n$, the number of linear equations exceeds the number of unknowns; therefore, we cannot expect to find any x such that $Ax = b$. In this case, we define a *least squares solution* x_{LS} to be any vector that satisfies

$$\|b - Ax_{LS}\|^2 = \min \|b - Ax\|^2, \tag{2.8}$$

where $\| \cdot \|$ is the Euclidean vector norm. In general, the least squares solution may be nonunique. In this case, the minimum norm solution can be generalized to mean the least squares solution with minimum norm. The error in the fit of the data, given by

$$r = b - Ax_{LS},$$

is called the *residual error* or *discrepancy* associated with x_{LS}.

In the case $m = n$, when there are as many linear equations as unknowns, there is guarantee that a unique solution exists for any choice of b only if the matrix A is invertible; otherwise, the solution of the linear systems will have to address the non-unicity and possibly non-existence of the solution. If some of the equations are redundant, i.e., they can be obtained as linear combinations of the other equations, they can be eliminated from the system, leading to an underdetermined system.

The choice of an algorithm for solving (2.4) often depends on several factors, the size of the problem being important, but not the only consideration. Algorithms for solving (2.4) can be broadly divided into the two classes of *direct methods* and *iterative methods*. In this chapter, we only provide a brief review of some direct methods. Iterative solvers that play an important role in Bayesian scientific computing will be presented in Chap. 9.

2.2.2 Direct Linear System Solvers

Direct methods solve (2.4) by first factorizing the matrix A into the product of matrices with special properties, then solving a sequence of simpler linear systems. Typically the factorization of the matrix A requires

(a) the availability of the matrix in explicit form;
(b) memory allocation for the matrix A and its factors;
(c) a number of floating point operations proportional to the cube of the dimension of the problem.

For these reasons, direct methods are most suitable for linear systems of small or medium dimensions, or for matrices with sparse representation. More importantly, for our scope, direct methods are of the "all or nothing" kind, in the sense that if the solution process is not completed, no partial information about the solution is retained. We are not going to present a systematic treatment of direct methods, a topic that is addressed with the due attention in standard linear algebra textbooks, but instead, we summarize the main algorithmic points through some examples.

Example 2.1 If (2.4) is a square system, i.e., $m = n$, a standard way to proceed to is to first factor the matrix into a product of a lower and an upper triangular matrix, and then to solve two triangular linear systems by back and forward substitution. A popular algorithm for performing the matrix factorization is Gaussian elimination, producing the *LU decomposition* of A,

$$A = LU,$$

where L is a lower triangular matrix with ones on the diagonal, and U is an upper triangular matrix,

$$L = \begin{bmatrix} 1 & & & \\ * & 1 & & \\ \vdots & & \ddots & \\ * & \cdots & * & 1 \end{bmatrix}, \quad U = \begin{bmatrix} * & * & \cdots & * \\ & * & & * \\ & & \ddots & \vdots \\ & & & * \end{bmatrix},$$

the asterisks representing possibly non-vanishing elements. We point out that in actual computations, row permutations of the matrix A may be needed to guarantee that the algorithm does not break down, see Notes and Comments at the end of the chapter. With the LU decomposition at hand, the original system is solved in two phases,

$$Ly = b, \quad Ux = y.$$

The first linear system is always solvable, while the solvability of the second one depends on the diagonal elements of U. If the diagonal of U contains zeros, U is singular and so is A. The LU decomposition is a built-in Matlab function, with the following syntax:

```
[L,U] = lu(A);
```

A square matrices $A \in \mathbb{R}^{n \times n}$ is *symmetric positive semi-definite* if it is symmetric, i.e., $A = A^{\mathsf{T}}$ and

$$v^{\mathsf{T}} A v \geq 0 \text{ for every } v \in \mathbb{R}^n.$$

A symmetric matrix A that satisfies the stronger condition

$$v^{\mathsf{T}} A v > 0 \text{ for every } v \in \mathbb{R}^n, \ v \neq 0.$$

is *symmetric positive definite* (SPD). Symmetric positive definite—and semi-definite—matrices play an important role in statistics.

While the symmetry of a matrix is immediate to verify, checking for positive definiteness based on the definition above is all but straightforward.

Consider the special case, in which the symmetric matrix A admits a factorization of the form

$$A = R^{\mathsf{T}} R, \quad R \in \mathbb{R}^{n \times n}. \tag{2.9}$$

Then for any vector $v \in \mathbb{R}^n$,

$$v^{\mathsf{T}} A v = \left(v^{\mathsf{T}} R^{\mathsf{T}}\right)\left(R v\right) = (R v)^{\mathsf{T}}(R v) = \|R v\|^2 \geq 0,$$

proving that the matrix is automatically symmetric positive semi-definite. If, in addition,

$$v^{\mathsf{T}} A v = \|R v\|^2 = 0 \text{ if and only if } v = 0,$$

which is equivalent to having $\mathcal{N}(R) = \{0\}$, and therefore R being invertible, the matrix A is SPD.

We have seen that all symmetric square matrices that admit a factorization (2.9) with R invertible are SPD. Conversely, it turns out that every SPD matrix admits a factorization of the form (2.9) with R invertible. One can prove that a symmetric matrix $A \in \mathbb{R}^{n \times n}$ is positive definite if and only if there exists a symmetric decomposition (2.9), where the matrix $R \in \mathbb{R}^{n \times n}$ is upper triangular,

$$R = \begin{bmatrix} r_{11} & r_{12} & \cdots & r_{1n} \\ & r_{22} & & r_{2n} \\ & & \ddots & \vdots \\ & & & r_{nn} \end{bmatrix},$$

the empty slots indicating zero entries, and, moreover, $r_{jj} > 0$ for all j, $1 \leq j \leq n$.

The matrix decomposition (2.9) where R is an invertible upper triangular matrix is the *Cholesky factorization* of A. Because of the easy availability of efficient numerical methods to compute the Cholesky factorization, often the most efficient way of figuring out whether a symmetric matrix is positive definite amounts to testing if the Cholesky factorization algorithm is successful.

The direct method of choice for solving a linear system of equations with an SPD matrix A is to compute the Cholesky factor R, and then solve the two linear systems

$$R^\mathsf{T} y = b, \quad Rx = y.$$

This requires only one half of the memory allocation needed for LU factorization, and the number of floating point operations is also reduced to about one half.

We conclude our discussion of symmetric positive definite matrices by looking at yet another characterization. A real symmetric matrix A admits an eigenvalue factorization of the form

$$A = Q^\mathsf{T} \Lambda Q, \tag{2.10}$$

where Q is an orthogonal matrix and Λ is a real diagonal matrix with diagonal entries $\lambda_j, 1 \le j \le n$. The columns of Q are mutually orthogonal eigenvectors of A, and the entries λ_j are the corresponding eigenvalues. The orthogonality of Q implies that

$$\|Qx\|^2 = (Qx)^\mathsf{T}(Qx) = \|x\|^2.$$

By defining $y = Qx$, we have

$$x^\mathsf{T} A x = (Qx)^\mathsf{T} \Lambda Qx = y^\mathsf{T} \Lambda y = \sum_{j=1}^{n} \lambda_j y_j^2.$$

If $\lambda_j > 0, \ 1 \le j \le n$, then for any $x \ne 0$, $x^\mathsf{T} A x > 0$, hence A is SPD.

Conversely, if A is SPD, then for any $x \ne 0$, $x^\mathsf{T} A x > 0$. Replacing A with its eigenvalue factorization (2.10) and letting $x = Q^\mathsf{T} e_j, \ 1 \le j \le n$, it follows that

$$(Q^\mathsf{T} e_j)^\mathsf{T} A (Q^\mathsf{T} e_j) = e_j^\mathsf{T} Q Q^\mathsf{T} \Lambda Q Q^\mathsf{T} e_j = e_j^\mathsf{T} \Lambda e_j = \lambda_j > 0, \quad 1 \le j \le n,$$

proving that a matrix is SPD if and only if it is symmetric and all its eigenvalues are real and positive.

Example 2.2 Consider an overdetermined linear system

$$Ax = b, \quad A = \begin{bmatrix} * & \cdots & * \\ * & \cdots & * \\ \vdots & & \vdots \\ * & \cdots & * \\ * & \cdots & * \end{bmatrix} \in \mathbb{R}^{m \times n}, \quad m > n,$$

where the solution is intended in the least squares sense.

Using the SVD of A, it can be shown that if the columns of A are linearly independent, the null space of A contains only the zero vector, hence the minimizer of (2.8) exists and is unique. To find the minimizer, x_{LS}, observe that the residual vector $r = b - Ax_{\text{LS}}$ must be orthogonal to the range of A, spanned by the column vectors in A, that is

$$A^\mathsf{T} r = A^\mathsf{T} (b - Ax_{\text{LS}}) = 0.$$

Therefore, the least squares solution must satisfy the *normal equations*,

$$A^T A x = A^T b, \tag{2.11}$$

where the matrix $A^T A \in \mathbb{R}^{n \times n}$ is symmetric and positive definite, since

$$v^T A^T A v = (Av)^T A v = \|Av\|^2 \geq 0,$$

with equality holding if and only if $Av = 0$. But since $\mathcal{N}(A) = \{0\}$, we have $v = 0$, showing that $A^T A$ is SPD. The normal equations can be solved by using the Cholesky factorization of $A^T A$.

Passing to the normal equations is not necessarily the best way to solve a least squares problem numerically, in particular when the coefficient matrix is ill-conditioned and the right-hand side is contaminated by noise. In fact, since the singular values of $A^T A$ are the squares of the nonzero singular values of A, the condition number of the normal equations is the square of the condition number of the original linear system, thus perturbations in the right hand side may become significantly emphasized. This issue is of particular concern when the condition number is large.

Rather than forming explicitly the normal equations, a standard way of solving numerically least squares problems relies on the *QR factorization* of the matrix A,

$$A = QR,$$

where $Q \in \mathbb{R}^{m \times m}$ is an orthogonal matrix, that is, $Q^{-1} = Q^T$ and R is upper triangular,

$$R = \begin{bmatrix} * & \cdots & * \\ & \ddots & \vdots \\ & & * \end{bmatrix} = \begin{bmatrix} R_1 \\ O_{(m-n) \times n} \end{bmatrix} \in \mathbb{R}^{m \times n}.$$

Multiplying both sides of (2.4) from the left by Q^T yields the triangular system

$$Rx = Q^T b,$$

which can be partitioned as

$$\begin{bmatrix} R_1 x \\ O_{(m-n) \times n} \end{bmatrix} = \begin{bmatrix} (Q^T b)_1 \\ (Q^T b)_2 \end{bmatrix} \begin{matrix} (n) \\ (m-n) \end{matrix}.$$

Whether the last $m - n$ equations are satisfied or not is entirely dependent on the right hand side b, while the first n equations are satisfied provided that the vector x is chosen so that

$$R_1 x = (Q^T b)_1.$$

Such a vector x can be found by back-substitution, if the diagonal elements of R_1 are nonzero, which is equivalent to the columns of A being linearly independent.

We remark that if the columns of A are linearly independent, the solution of the normal equations is

$$x_{\text{LS}} = \left(A^{\mathsf{T}}A\right)^{-1}A^{\mathsf{T}}b = A^{\dagger}b.$$

The matrix $A^{\dagger} = \left(A^{\mathsf{T}}A\right)^{-1}A^{\mathsf{T}} \in \mathbb{R}^{n \times m}$ is called the *pseudoinverse* of A. It is easy to check that if A is square and invertible, the pseudoinverse coincides with the usual inverse.

Often, the pseudoinverse is more of theoretical than computational interest, and it can be helpful tool when analyzing the solution of linear systems with non-invertible matrices.

Notes and Comments

There is a wealth of textbooks in linear algebra to choose from, see, e.g., [21, 45, 71]. For a more computational approach, we refer to [80].

We point out here that the discussion of the LU decomposition, as mentioned in Example 2.1, is slightly streamlined, as it omits the often necessary row permutations that constitute part of the decomposition, a process known as the partial pivoting in Gaussian elimination. For details, we refer to the cited literature.

Chapter 3
Continuous and Discrete Multivariate Distributions

After the generic introduction to probability and linear algebra, in this chapter, we start to put together these concepts. In particular, we introduce the workhorse of computational statistics, the normal distribution, using elements from matrix factorizations. Normal distributions play a role in computational statistics similar to that of linear operators in analysis local linearizations of non-linear mappings, crucial for designing efficient computational algorithms, have the counterpart of normal approximations in computational statistics. Furthermore, in anticipation of sampling methods, we also discuss discrete distributions, and in particular, the Poisson distribution that has a central role in modeling rare events.

3.1 Covariance Matrices

The importance of positive definiteness in statistics is clear when considering covariance matrices. Let X denote an n-variate random variable, $X : \Omega \to \mathbb{R}^n$, and assume that the mean and covariance of X,

$$\overline{x} = \mathrm{E}(X), \quad \mathsf{C} = \mathrm{E}\big((X - \overline{x})(X - \overline{x})^{\mathsf{T}}\big),$$

are well-defined. The variances of the components X_i, encoded in the diagonal entries c_{ii} of the covariance matrix C, measure the variability of the random variable around the mean along the coordinate directions. However, the coordinate directions need not play any particular role among all directions, and we may be interested in the variability in other directions. Given the vector $v \neq 0$ in $\mathbb{R}^n$, define the real-valued random variable

$$X_v = v^{\mathsf{T}} X = \sum_{i=1}^{n} v_i X_i,$$

D. Calvetti and E. Somersalo, *Bayesian Scientific Computing*, Applied Mathematical Sciences 215, https://doi.org/10.1007/978-3-031-23824-6_3

whose mean and variance can be calculated as

$$E(X_v) = \int_{\mathbb{R}^n} v^{\mathsf{T}} x\, \pi_X(x) dx = v^{\mathsf{T}} \left(\int_{\mathbb{R}^n} x\, \pi_X(x) dx \right) = v^{\mathsf{T}} \bar{x},$$

and

$$\begin{aligned}
\mathrm{var}(X_v) &= \int_{\mathbb{R}^n} (v^{\mathsf{T}} x - v^{\mathsf{T}} \bar{x})^2\, \pi_X(x) dx \\
&= \int_{\mathbb{R}^n} \left(v^{\mathsf{T}} (x - \bar{x}) \right)^2 \pi_X(x) dx \\
&= \int_{\mathbb{R}^n} v^{\mathsf{T}} (x - \bar{x})(x - \bar{x})^{\mathsf{T}} v\, \pi_X(x) dx \\
&= v^{\mathsf{T}} \left(\int_{\mathbb{R}^n} (x - \bar{x})(x - \bar{x})^{\mathsf{T}} \pi_X(x) dx \right) v \\
&= v^{\mathsf{T}} \mathsf{C} v.
\end{aligned}$$

By definition, the variance of a real-valued random variable is non-negative, implying, in particular, that

$$v^{\mathsf{T}} \mathsf{C} v \geq 0.$$

Moreover, since the covariance matrix is symmetric, we conclude that covariance matrices are always symmetric positive semidefinite. A legitimate question to ask is whether we can expect, in general, covariance matrices to be SPD. The answer is negative, as can be understood by a simple example: Let $X_1 : \Omega \to \mathbb{R}$ be a real-valued random variable with vanishing mean and finite variance $\sigma^2 > 0$, and define $X_2 : \Omega \to \mathbb{R}$ as $X_2(\omega) = 0$. The random variable

$$X = \begin{bmatrix} X_1 \\ X_2 \end{bmatrix},$$

has covariance matrix

$$\mathsf{C} = \begin{bmatrix} \sigma_1^2 & 0 \\ 0 & 0 \end{bmatrix},$$

which is not SPD, since

$$\begin{bmatrix} 0 \\ 1 \end{bmatrix} \mathsf{C} \begin{bmatrix} 0 & 1 \end{bmatrix} = 0,$$

that is, the random variable X has no variability in the second coordinate direction. In this case, we say that the random variable X is *degenerate*. Observe that the second coordinate direction is aligned with the null space of the covariance matrix. Therefore, by restricting the random variable X to the orthocomplement of the null space, coinciding with $\mathcal{R}(\mathsf{C}^{\mathsf{T}}) = \mathcal{R}(\mathsf{C})$, the degeneracy disappears.

3.2 Normal Distributions

The normal, or Gaussian distribution, is the workhorse of Bayesian scientific computing and deserves to be discussed in detail. As usual, we start by addressing the one-dimensional case, then consider the general multivariate case.

A random variable $X \in \mathbb{R}$ is *normally distributed*, or Gaussian, indicated symbolically by

$$X \sim \mathcal{N}(\mu, \sigma^2),$$

if its cumulative distribution is given by

$$P\{X \le t\} = \frac{1}{\sqrt{2\pi\sigma^2}} \int_{-\infty}^{t} \exp\left(-\frac{1}{2\sigma^2}(x-\mu)^2\right) dx.$$

Hence, the Gaussian probability density is

$$\pi_X(x) = \frac{1}{\sqrt{2\pi\sigma^2}} \exp\left(-\frac{1}{2\sigma^2}(x-\mu)^2\right).$$

It is a straightforward integration exercise to show that

$$E(X) = \mu, \quad \text{var}(X) = \sigma^2.$$

We are now ready to introduce multivariate normal distributions. Given n mutually independent Gaussian random variables X_j,

$$X_j \sim \mathcal{N}(\mu_j, \sigma_j^2), \quad 1 \le j \le n,$$

a multivariate Gaussian random variable X with components X_j has the probability density

$$\pi_X(x) = \pi_{X_1}(x_1) \cdots \pi_{X_n}(x_n)$$

$$= \left(\frac{1}{(2\pi)^n \sigma_1^2 \cdots \sigma_n^2}\right)^{1/2} \exp\left(-\frac{1}{2} \sum_{j=1}^{n} \frac{(x_j - \mu_j)^2}{\sigma_j^2}\right). \tag{3.1}$$

This formula leads to a natural generalization of the multivariate Gaussian distribution for non-independent components. Introduce the invertible diagonal matrix $D \in \mathbb{R}^{n \times n}$,

$$D = \begin{bmatrix} \sigma_1^2 & & \\ & \ddots & \\ & & \sigma_n^2 \end{bmatrix},$$

whose inverse is

$$\mathbf{D}^{-1} = \begin{bmatrix} 1/\sigma_1^2 & & \\ & \ddots & \\ & & 1/\sigma_n^2 \end{bmatrix}.$$

For any $z \in \mathbb{R}^n$, we have

$$z^{\mathsf{T}}\mathbf{D}^{-1}z = \sum_{j=1}^{n} \frac{z_j^2}{\sigma_j^2}, \quad \det(\mathbf{D}) = \sigma_1^2\sigma_2^2\cdots\sigma_n^2,$$

therefore, letting $z = x - \mu$ and substituting in (3.1), we get

$$\pi_X(x) = \left(\frac{1}{(2\pi)^n \det(\mathbf{D})}\right)^{1/2} \exp\left(-\frac{1}{2}(x-\mu)^{\mathsf{T}}\mathbf{D}^{-1}(x-\mu)\right),$$

where $\mu \in \mathbb{R}^n$ is the vector with components μ_j, $1 \le j \le n$.

To understand the role of independency of the components of X, observe that since $\mathbf{D}$ is diagonal, the sets

$$(x-\mu)^{\mathsf{T}}\mathbf{D}^{-1}(x-\mu) = \text{constant}$$

represent ellipsoidal hypersurfaces in $\mathbb{R}^n$ with the principal axes parallel to the coordinate axes. We may therefore ask what happens if we rotate the coordinate system around the point μ so that the principal axes are no longer parallel to the coordinate axes. Let $\mathbf{U} \in \mathbb{R}^{n \times n}$ denote a rotation matrix in $\mathbb{R}^n$. Define the new random variable Z,

$$Z = \mu + \mathbf{U}(X - \mu),$$

obtained by first centering the random vector X at the origin by subtracting its mean, rotating, and, finally, moving it back around the vector μ. Recalling that a rotation matrix is orthogonal, hence its inverse coincides with its transpose, we obtain the identity

$$X - \mu = \mathbf{U}^{\mathsf{T}}(Z - \mu),$$

thus

$$X = \mu + \mathbf{U}^{\mathsf{T}}(Z - \mu) = \phi(Z).$$

Since Z is the result of a differentiable one-to-one coordinate transformation, we can apply the procedure outlined in Chap. 1 to find its probability density. First, notice that the Jacobian matrix of ϕ is

$$D_z\phi = \mathbf{U}^{\mathsf{T}},$$

and since U is orthogonal, the determinant of the Jacobian is

$$\frac{\partial \phi}{\partial z} = \det(U^{\mathsf{T}}) = \pm 1.$$

Furthermore, the substitution $x - \mu = U^{\mathsf{T}}(z - \mu)$ in the probability density function of X yields

$$(x - \mu)^{\mathsf{T}} D^{-1}(x - \mu) = (z - \mu)^{\mathsf{T}} U D^{-1} U^{\mathsf{T}}(z - \mu) = (z - \mu)^{\mathsf{T}}(U D U^{\mathsf{T}})^{-1}(z - \mu),$$

where we used the fact that $U^{\mathsf{T}} = U^{-1}$, hence

$$(U D U^{\mathsf{T}})^{-1} = (U^{\mathsf{T}})^{-1} D^{-1} U^{-1} = U D^{-1} U^{\mathsf{T}}.$$

Introducing the matrix $C \in \mathbb{R}^{n \times n}$,

$$C = U D U^{\mathsf{T}}, \tag{3.2}$$

we observe that

$$\det(C) = \det(U D U^{\mathsf{T}}) = \det(U)\det(D)\det(U^{\mathsf{T}})$$
$$= \left(\det(U)\right)^2 \det(D) = \det(D).$$

Therefore, we conclude that the probability density of the rotated variable Z is given by

$$\pi_Z(z) = \left(\frac{1}{(2\pi)^n \det(C)}\right)^{1/2} \exp\left(-\frac{1}{2}(z - \mu)^{\mathsf{T}} C^{-1}(z - \mu)\right), \tag{3.3}$$

where C is a symmetric positive definite matrix. We can verify through an n-dimensional integration argument that

$$E(Z) = \mu, \quad \mathrm{cov}(Z) = C.$$

This way, we have defined the general Gaussian multivariate random variable, and we indicate this by writing

$$Z \sim \mathcal{N}(\mu, C).$$

Later on, we write occasionally the probability density of a Gaussian, or normally distributed random variable as

$$\pi_Z(z) = \mathcal{N}(z \mid \mu, C).$$

A *standard normal* random variable is a Gaussian random variable with zero mean, $\mu = 0$, and covariance the $n \times n$ identity matrix, i.e., $C = I_n$.

3.3   How Normal is it to be Normal?

Normal densities are often used for convenience, as the two parameters, mean and covariance, that completely characterize them, provide an intuitive interpretation of the density. More generally, Gaussian random variables model well *macroscopic* measurements that are sums of individual *microscopic* random effects, such as in the case of pressure, temperature, electric current, and luminosity. The ubiquity of Gaussian random variables can be understood and justified in the light of the Central Limit Theorem, stated here for completeness. A proof can be found in standard textbooks, see Notes and Comments at the end of the chapter.

Central Limit Theorem Assume that the real-valued random variables $X_1, X_2, \ldots$ are *independent* and *identically distributed* (i.i.d.), that is, the probability distributions of X_j are equal. Assume further that the common probability distribution has finite mean and variance, denoted by $\mu \in \mathbb{R}$ and $\sigma^2 > 0$, respectively. Then the probability distribution of the random variable

$$Z_n = \frac{1}{\sigma\sqrt{n}}(X_1 + X_2 + \cdots + X_n - n\mu), \quad n = 1, 2, \ldots$$

converges to the distribution of a standard normal random variable in the sense that

$$\lim_{n\to\infty} P\{Z_n \le x\} = \frac{1}{2\pi} \int_{-\infty}^{x} e^{-t^2/2} dt.$$

Observe that

$$E(Z_n) = \frac{1}{\sigma\sqrt{n}}\left(\sum_{j=1}^{n} E(X_j) - n\mu\right) = 0,$$

and the scaling of the sum is chosen so that

$$\mathrm{Var}(Z_n) = E(Z_n^2)$$

$$= \frac{1}{n\sigma^2}E\left(\left(\sum_{j=1}^{n}(X_j - \mu)\right)^2\right)$$

$$= \frac{1}{n\sigma^2}E\left(\sum_{j=1}^{n}(X_j - \mu)^2 + 2\sum_{j>\ell}(X_j - \mu)(X_\ell - \mu)\right).$$

By the mutual independence of the zero mean random variables $(X_j - \mu)$, the contribution of the double sum vanishes, since

$$E\left(\sum_{j>\ell}(X_j - \mu)(X_\ell - \mu)\right) = \sum_{j>\ell} E(X_j - \mu)E(X_\ell - \mu) = 0,$$

and therefore

$$\text{Var}(Z_n) = \frac{1}{n\sigma^2} \sum_{j=1}^{n} \text{E}(X_j - \mu)^2 = 1,$$

indicating why one can expect the limit to have zero mean and unit variance.

Another way of thinking about the result of the Central Limit Theorem is the following. Let Y_n denote the average of n identically distributed independent random variables with mean μ and variance σ^2,

$$Y_n = \frac{1}{n} \sum_{j=1}^{n} X_j.$$

We observe that

$$\text{E}(Y_n) = \frac{1}{n} \sum_{j=1}^{n} \text{E}(X_j) = \mu,$$

and by writing

$$Y_n - \mu = \frac{1}{n} \left(\sum_{j=1}^{n} X_j - n\mu \right) = \frac{\sigma}{\sqrt{n}} Z_n,$$

we conclude that the variance of Y_n is

$$\text{Var}(Y_n) = \text{E}\big((Y_n - \mu)^2\big) = \frac{\sigma^2}{n}.$$

The Central Limit Theorem states that, for n large, a good approximation for the probability distribution of Y_n is

$$Y_n \sim \mathcal{N}\left(\mu, \frac{\sigma^2}{n}\right),$$

thus confirming the intuitive idea that by averaging random variables with common mean, in the limit we arrive to that mean.

3.4 Discrete Distributions

In the discussion so far, we have assumed the existence of a probability density which is a non-negative function. In the case where the random variable can take on only discrete values, we need to introduce singular point masses to define its distribution. In the following, we build a bridge between discrete and continuous densities that later will be crucial to understand how probability distributions can be approximated using sampling techniques.

A point mass in $\mathbb{R}$ is a measure that is concentrated at a single point. The concept of point mass can be derived from normal distributions via a limiting process as follows. We start with a random variable Y following the standard normal distribution:

$$Y \sim \mathcal{N}(0, 1), \quad \text{or} \quad \pi_Y(y) = \frac{1}{\sqrt{2\pi}} e^{-y^2/2} = \psi(y).$$

For $x_0 \in \mathbb{R}$ and $\varepsilon > 0$, we define a new random variable X through an affine transformation

$$X = x_0 + \varepsilon Y \Leftrightarrow Y = \frac{X - x_0}{\varepsilon} = \varphi(X).$$

Using the change of variables formula for probability densities, we find that the density of X is

$$\pi_X(x) = \pi_Y(\varphi(X)) \frac{d\varphi}{dx}(x) = \frac{1}{\sqrt{2\pi}\varepsilon} e^{-(x-x_0)^2/2\varepsilon^2} = \frac{1}{\varepsilon} \psi\left(\frac{x - x_0}{\varepsilon}\right).$$

We observe that X is a Gaussian random variable centered at x_0 and variance ε^2. Next, we investigate what happens when ε tends to zero. Intuitively, the probability mass is more and more concentrated around x_0. It is not difficult to verify that, pointwise, the probability density converges to zero at every point except for x_0, where the point value tends to infinity. Since the pointwise limit is not particularly useful, we formulate our question in a different way. Let f be any bounded continuous function. We want to find out what happens to the integral

$$I_\varepsilon = \int_{\mathbb{R}} f(x)\pi_X(x)dx = \frac{1}{\varepsilon} \int_{\mathbb{R}} f(x)\psi\left(\frac{x - x_0}{\varepsilon}\right)dx,$$

as $\varepsilon \to 0+$. To answer this question, we change back to the variable y, by making the substitution

$$x = x_0 + \varepsilon y, \quad dx = \varepsilon \, dy,$$

and observe that, if

$$I_\varepsilon = \int_{\mathbb{R}} f(x_0 + \varepsilon y)\psi(y)dy,$$

then

$$\lim_{\varepsilon \to 0+} I_\varepsilon = f(x_0) \int_{\mathbb{R}} \psi(y)dy = f(x_0).$$

This calculation justifies the following characterization. A *point mass function* at $x_0 \in \mathbb{R}$, denoted by $\delta_{x_0}(x)$ is characterized by the property that for all bounded continuous functions f,

$$\int_{\mathbb{R}} f(x)\delta_{x_0}(x)dx = f(x_0),$$

or in other words, the point mass defines a point evaluation functional,

$$\delta_{x_0} : f \mapsto f(x_0).$$

In particular, for $f(x) = 1$ for all x, we have

$$\int_{\mathbb{R}} \delta_{x_0}(x)dx = 1.$$

A point mass function is also referred to as Dirac's delta function. Clearly, a point mass is not a classical function but an instance of a generalized function.

The introduction of a point mass function allows us to define the large class of random variables taking on values in a discrete set. Consider a random variable X that assumes only non-negative integer values $j = 0, 1, 2, \ldots$, and let

$$p_j = P\{X = j\}.$$

Define the probability distribution of X as

$$\pi_X(x) = \sum_{j=0}^{\infty} p_j \delta_j(x).$$

where the p_js satisfy

$$\int_{\mathbb{R}} \pi_X(x)dx = \sum_{j=0}^{\infty} p_j = 1.$$

We then compute the expected value of X,

$$\bar{x} = E(X) = \int_{\mathbb{R}} x\pi_X(x)dx = \sum_{j=1}^{\infty} p_j \int_{\mathbb{R}} x\delta_j(x)dx = \sum_{j=0}^{\infty} jp_j,$$

and observe that there is no guarantee that it is an integer. In that sense, the term expected value may be a misnomer, as we may not expect X to take on that value at all. Similarly, the variance of X is given by

$$\mathrm{var}(X) = \int_{R} (x - \bar{x})^2 \pi_X(x)dx = \sum_{j=0}^{\infty} (j - \bar{x})^2 p_j.$$

A popular discrete distribution to model rare events is the *Poisson distribution*. Before formally defining it, we present the following motivating example.

Example 3.1 This example deals with modeling photon counts. A weak light source emits photons that are counted with a *charged coupled device* (CCD). The *counting process*

$$N(t) = \{\text{number of particles observed in } [0, t]\} \in \mathbb{N}$$

is an integer-valued random variable that depends on the parameter $t > 0$.

To set up a statistical model for $N(t)$, we make the following assumptions:

1. *Stationarity:* Let Δ_1 and Δ_2 be any two time intervals of equal length, and let n be a non-negative integer. Then

$$P\{n \text{ photons arrive in } \Delta_1\} = P\{n \text{ photons arrive in } \Delta_2\}.$$

2. *Independent increments:* Let $\Delta_1, \ldots, \Delta_n$ be non-overlapping time intervals, let $k_1, \ldots, k_n$ be non-negative integers, and let A_j denote the event

$$A_j = \{k_j \text{ photons arrive in the time interval } \Delta_j\}.$$

Then the events $A_1, \cdots, A_n$ are mutually independent, i.e.,

$$\mathbb{P}(A_1 \cap \cdots \cap A_n) = \mathbb{P}(A_1) \cdots \mathbb{P}(A_n).$$

3. *Negligible probability of coincidence:* Assume that the probability of two or more events occurring at the same time is negligible. More precisely, assume that $N(0) = 0$ and

$$\lim_{h \to 0} \frac{P\{N(h) > 1\}}{h} = 0.$$

This condition can be interpreted by saying that the number of counts increases at most linearly.

If these assumptions hold, then it can be shown (see Notes and Comments for a reference) that N is a *Poisson process* and

$$P\{N(t) = j\} = \frac{(\lambda t)^j}{j!} e^{-\lambda t}, \quad \lambda > 0.$$

The previous example motivates us to define the random variable X with discrete probability density,

$$\pi_X(x) = \sum_{j=0}^{\infty} p_j \delta_j(x), \quad p_j = \frac{\theta^j}{j!} e^{-\theta}, \tag{3.4}$$

where $\theta > 0$ is the *Poisson parameter*. We write

$$X \sim \text{Poisson}(\theta).$$

The calculation of the expectation of this Poisson random variable is rather straight-forward:

$$E(X) = e^{-\theta} \sum_{j=0}^{\infty} j \frac{\theta^j}{j!}$$

$$= e^{-\theta} \sum_{j=1}^{\infty} \frac{\theta^j}{(j-1)!} = e^{-\theta} \sum_{j=0}^{\infty} \frac{\theta^{j+1}}{j!}$$

$$= \theta e^{-\theta} \underbrace{\sum_{j=0}^{\infty} \frac{\theta^j}{j!}}_{=e^{\theta}} = \theta.$$

To calculate the variance of a Poisson random variable, we first observe that

$$E((X - \theta)^2) = E(X^2) - 2\theta \underbrace{E(X)}_{=\theta} + \theta^2$$

$$= E(X^2) - \theta^2$$

$$= \sum_{j=0}^{\infty} j^2 p_j - \theta^2.$$

Substituting the expression for p_j from (3.4) yields

$$E((X - \theta)^2) = e^{-\theta} \sum_{j=0}^{\infty} j^2 \frac{\theta^j}{j!} - \theta^2 = e^{-\theta} \sum_{j=1}^{\infty} j \frac{\theta^j}{(j-1)!} - \theta^2$$

$$= e^{-\theta} \sum_{j=0}^{\infty} (j+1) \frac{\theta^{j+1}}{j!} - \theta^2$$

$$= \theta e^{-\theta} \left\{ \sum_{j=0}^{\infty} j \frac{\theta^j}{j!} + \sum_{j=0}^{\infty} \frac{\theta^j}{j!} \right\} - \theta^2$$

$$= \theta e^{-\theta} (\theta e^{\theta} + e^{\theta}) - \theta^2$$

$$= \theta.$$

Thus, the mean and the variance of the Poisson random variable X are equal to the parameter characterizing the distribution.

3.4.1 Normal Approximation to the Poisson Distribution

We now return to the example of the CCD camera, and notice that the total photon count can be thought of as a sum of random events, single photon arrivals, that are independent and identically distributed. Therefore, it is reasonable to ask if the Poisson distribution can be reliably approximated by a normal distribution when the count number is high. In order to show that this is a reasonable approximation, we need to introduce some auxiliary results.

Given two mutually independent random variables, both following a Poisson distribution:

$$X_j \sim \text{Poisson}(\theta_j), \quad j = 1, 2,$$

their sum is also Poisson distributed, and

$$X = X_1 + X_2 \sim \text{Poisson}(\theta_1 + \theta_2).$$

To prove the claim, we write

$$P\{X = k\} = P\{(X_1, X_2) = (k, 0) \text{ or } (X_1, X_2) = (k - 1, 1) \text{ or } \\ \ldots \text{ or } (X_1, X_2) = (0, k)\}$$

$$= P\{(X_1, X_2) = (k, 0)\} + P\{(X_1, X_2) = (k - 1, 1)\} + \\ \ldots + P\{(X_1, X_2) = (0, k)\}.$$

Further, by the independency of X_1 and X_2,

$$P\{(X_1, X_2) = (k - j, j)\} = P\{X_1 = k - j\}P\{X_2 = j\} = e^{-(\theta_1 + \theta_2)} \frac{\theta_1^{k-j}}{(k - j)!} \frac{\theta_2^{j}}{j!},$$

which implies that

$$P\{X = k\} = e^{-(\theta_1 + \theta_2)} \sum_{j=0}^{k} \frac{\theta_1^{k-j}}{(k - j)!} \frac{\theta_2^{j}}{j!} = \frac{1}{k!} e^{-(\theta_1 + \theta_2)} \sum_{j=0}^{k} \binom{k}{j} \theta_1^{k-j} \theta_2^{k}$$

$$= \frac{(\theta_1 + \theta_2)^k}{k!} e^{-(\theta_1 + \theta_2)}$$

by the binomial formula, thus completing the proof.

Consider now a Poisson distributed random variable, and assume, for simplicity, that the Poisson parameter θ is an integer. We remark that this is not a necessary requirement, but a way to simplify the discussion. We can represent X in an equivalent form as

$$X = \sum_{j=1}^{\theta} X_j, \quad X_j \sim \text{Poisson}(1),$$

where the variables X_j are assumed to be mutually independent. By induction, the reasoning above shows that the sum is indeed a Poisson-distributed random variable. By the Central Limit Theorem, we conclude that, approximately,

$$\frac{X - \theta}{\sqrt{\theta}} \sim \mathcal{N}(0, 1),$$

and the approximation becomes more accurate when θ grows. In particular, for all $a, b, 0 \le a < b$, we have the approximation

$$\sum_{a < j \le b} \frac{\theta^j}{j!} e^{-\theta} \approx \frac{1}{\sqrt{2\pi\theta}} \int_a^b e^{-\frac{1}{2\theta}(x-\theta)^2} dx,$$

establishing the normal approximation.

It turns out that the above approximation of a Poisson random distribution with a Gaussian can be slightly improved. If n is an integer, the probabilities $P\{X \le n + \delta\}$ coincide for all $\delta, 0 \le \delta < 1$. Therefore, the normal approximation becomes ambiguous, since

$$\sum_{j \le n} \frac{\theta^j}{j!} e^{-\theta} = \sum_{j \le n+\delta} \frac{\theta^j}{j!} e^{-\theta} \approx \frac{1}{\sqrt{2\pi\theta}} \int_{-\infty}^{n+\delta} e^{-\frac{1}{2\theta}(x-\theta)^2} dx, \quad 0 \le \delta < 1.$$

In practice, the value $\delta = 1/2$ is used, with this shift of the upper bound referred to as *continuity correction*.

Observe that if Y is the random variable corresponding to the normal density with continuity correction, then Y has the probability density

$$\pi_Y(y) = \frac{d}{dy} \left(\frac{1}{\sqrt{2\pi\theta}} \int_{-\infty}^{y+1/2} e^{-\frac{1}{2\theta}(x-\theta)^2} dx \right) = \frac{1}{\sqrt{2\pi\theta}} e^{-\frac{1}{2\theta}(y-\theta+1/2)^2},$$

that is,

$$Y \sim \mathcal{N}(\theta - 1/2, \theta).$$

To get an idea of the quality of the approximation, we plot the Gaussian $\mathcal{N}(\theta - 1/2, \theta)$ versus the Poisson probabilities p_j with parameter θ. Fig. 3.1 shows the plots for different values of θ.

The Central Limit Theorem justifies in many circumstances the use of Gaussian approximations. As we shall see, normal distributions are computationally quite convenient.

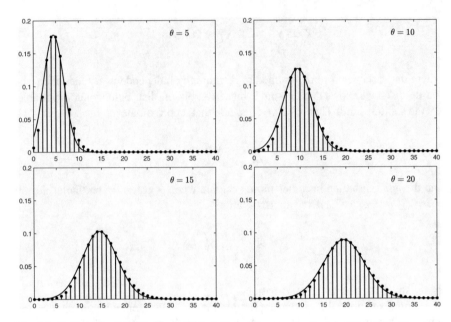

Fig. 3.1 Poisson distributions (*dots*) and their Gaussian approximations (*solid curve*) with various values of the mean θ

Notes and Comments

For the proof and discussion of the Central Limit Theorem, we refer to [5]. The proof concerning the statement about the Poisson distribution can be found, e.g., in [35].

The discovery of the normal distribution is often attributed to Abraham deMoivre, while systematically developed by Carl Friedrich Gauss and Pierre-Simon Laplace. A name worth mentioning is Adolphe Quetelet, the nineteenth-century Belgian statistician who in his famous 1835 treatise *Sur l'homme et le développement des ses facultés, ou, Essay de physique sociale,* introduced "l'homme moyen", the average man, bringing the Gaussian bell curve to anthropometric social sciences, and is considered as one of the founders of sociology, originally also known as social physics.

Chapter 4
Introduction to Sampling

A large portion of statistical inference deals with statistical modeling and analysis, which often include the description and understanding of the data collection process. Therefore, statistical inference lies at the very core of scientific modeling and empirical testing of theories. Statistical modeling may lead to the conclusion that the underlying probability density can be described in a parametric form, such as a Gaussian distribution, and the problem is reduced to one of estimating the parameters that characterize it. Often, however, there are not enough reasons to assume a parametric model, and one needs to resort to non-parametric models. In either case, it may be useful to consider summary statistics of the sample, such as mean and variances, as a first step. As discussed in this section, these concepts of descriptive statistics lead to more sophisticated ideas that turn out to be useful, for instance, when statistical methods are used in computational inverse problems.

4.1 On Averaging

The most elementary way to summarize a large set of statistical data is to calculate its average. Assuming that the data consist of a large sample of vectors,

$$S = \{x^1, x^2, \ldots, x^N\},$$

that are realizations from an underlying probability density π_X of a random variable X, a natural approximation of the mean of this putative density is

$$E(X) = \int x \, \pi_X(x) dx \approx \frac{1}{N} \sum_{j=1}^{N} x^j.$$

© The Author(s), under exclusive license to Springer Nature Switzerland AG 2023
D. Calvetti and E. Somersalo, *Bayesian Scientific Computing*, Applied Mathematical
Sciences 215, https://doi.org/10.1007/978-3-031-23824-6_4

The rationale and theoretical justification of this approximation is provided by the Law of Large Numbers. This law has several variants, one of which is the following.

Law of Large Numbers: Assume that $X^1, X^2, \ldots$ are independent and identically distributed random variables,

$$X^j \sim \pi_X,$$

with finite mean μ. Then

$$\lim_{N \to \infty} \frac{1}{N}(X^1 + X^2 + \cdots + X^N) = \mu$$

almost certainly, that is,

$$P\Big\{ \lim_{N \to \infty} \frac{1}{N} \sum_{j=1}^{N} X^j = \mu \Big\} = 1.$$

This formulation of the Law of Large Numbers is often referred to as the strong law of large numbers. The expression *almost certainly* means, in particular, that, with probability one, the averages of any realizations $x^1, x^2, \ldots$ of the random variables $X^1, X^2, \ldots$ converge towards the mean. This is good news, since data sets are always realizations. Observe the close connection with the Central Limit Theorem, which states the convergence in terms of probability densities.

More generally, we may want to calculate estimators for the expectation of a known function $f(X)$ based on a sample of realizations. Observing that if $X^1, \ldots, X^N$ are independent and identically distributed, then so are the random variables $f(X^1), \ldots, f(X^N)$, and the Law of Large Numbers justifies the approximation

$$E\big(f(X)\big) \approx \frac{1}{M} \sum_{j=1}^{N} f(x^j), \tag{4.1}$$

assuming that the expectation on the left is finite.

The following example is meant to clarify the strength of a parametric model if such model is available, and to what extent the simple summary statistics may be a misleading way to describe a probability distribution.

Example 4.1 Given a sample of points in the plane,

$$S = \big\{x^1, x^2, \ldots, x^N\big\}, \quad x^j = \begin{bmatrix} x_1^j \\ x_2^j \end{bmatrix} \in \mathbb{R}^2,$$

we want to set up a parametric model for the data, by assuming that the points are independent realizations of a normally distributed random variable,

$$X \sim \mathcal{N}(\mu, \mathbf{C}), \tag{4.2}$$

with unknown mean $\mu \in \mathbb{R}^2$ and covariance matrix $\mathbf{C} \in \mathbb{R}^{2\times 2}$. For future reference, we express this by writing the probability density of X using the notation of conditional densities,

$$\pi_X(x \mid \mu, \mathbf{C}) = \frac{1}{2\pi \det(\mathbf{C})^{1/2}} \exp\left(-\frac{1}{2}(x - \mu)^{\mathsf{T}}\mathbf{C}^{-1}(x - \mu)\right),$$

i.e., this is the form of the probability density *provided that μ and $\mathbf{C}$ are given*. The inference problem is therefore to estimate the parameters μ and $\mathbf{C}$ from the sample S.

As the Law of Large Numbers suggests, we may hope that N is sufficiently large to justify an approximation of the form

$$\mu = \mathrm{E}(X) \approx \frac{1}{N}\sum_{j=1}^{N} x^j = \widehat{\mu}. \tag{4.3}$$

More precisely, above we have assumed that we have N independent copies X^j, $1 \le j \le N$, of identically distributed independent random variables with the same distribution as X, and x^j is a realization of X^j.

To estimate the covariance matrix $\mathbf{C} = \mathrm{cov}(X) = \mathrm{E}\big((X - \mu)(X - \mu)^{\mathsf{T}}\big)$, we define the matrix valued function

$$f : \mathbb{R}^2 \to \mathbb{R}^{2\times 2}, \quad f(x) = (x - \widehat{\mu})(x - \widehat{\mu})^{\mathsf{T}},$$

and, combining it with (4.1), arrive at the approximation

$$\mathbf{C} \approx \widehat{\mathbf{C}} = \mathrm{E}\big(f(X)\big) \approx \frac{1}{N}\sum_{j=1}^{N}(x^j - \widehat{\mu})(x^j - \widehat{\mu})^{\mathsf{T}} \tag{4.4}$$

The estimates $\widehat{\mu}$ and $\widehat{\mathbf{C}}$ of the mean and covariance, respectively, from formulas (4.3) and (4.4), are often referred to as the *empirical mean and covariance*, or *sample mean and covariance*, respectively.

Before illustrating these ideas with examples, a comment is in order. The approximation (4.4) is often referred to as a biased estimator of the covariance matrix. In the statistics literature, the preferred covariance estimate is a slightly different approximation,

$$\widehat{\mathbf{C}}_u = \frac{1}{N-1}\sum_{j=1}^{N}(x^j - \widehat{\mu})(x^j - \widehat{\mu})^{\mathsf{T}} = \frac{N}{N-1}\widehat{\mathbf{C}}, \tag{4.5}$$

known as the *unbiased covariance estimator*. The reason for the different scaling and the terminology is explained in the Notes and Comments at the end of the chapter. Clearly, for N large, the difference between the estimators becomes insignificant.

We are now ready to present some computed examples. In the first one, we assume an underlying Gaussian parametric model that explains the sample, which was in fact drawn from a two-dimensional Gaussian density. The left panel of Fig. 4.1 shows the equiprobability curves of the original Gaussian probability density, which are ellipses. A scatter plot of a sample of $N = 500$ points drawn from this density is shown in the right panel of the same figure. We then compute the eigenvalue decomposition of the empirical covariance matrix,

$$\widehat{C} = UDU^T, \tag{4.6}$$

where $U \in \mathbb{R}^{2\times2}$ is an orthogonal matrix whose columns are the eigenvectors scaled to have unit length, and $D \in \mathbb{R}^{2\times2}$ is the diagonal matrix with the diagonal entries equal to the eigenvalues of $\widehat{C}$,

$$U = \begin{bmatrix} v^{(1)} & v^{(2)} \end{bmatrix}, \quad D = \begin{bmatrix} \lambda_1 & \\ & \lambda_2 \end{bmatrix}, \quad \widehat{C}v^{(j)} = \lambda_j v^{(j)}, \quad j = 1, 2.$$

The right panel of Fig. 4.1 shows the eigenvectors of $\widehat{C}$ scaled to have length twice the square root of the corresponding eigenvalues, applied at the empirical mean. The plot is in good agreement with our expectation.

To better understand the implications of adopting a parametric form for the underlying density, we now consider a sample from a density that is far from Gaussian. The equiprobability curves of this non-Gaussian density are shown in the left panel of Fig. 4.2. The right panel shows the scatter plot of a sample of $N = 500$ points drawn from this density. We compute the empirical mean, the empirical covariance matrix, and its eigenvalue decomposition. As in the Gaussian case, we plot the eigenvectors scaled by a factor twice the square root of the corresponding eigenvalues, applied at

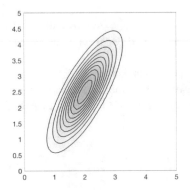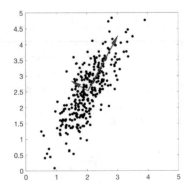

Fig. 4.1 The ellipses of the equiprobability curves of the original Gaussian distribution (left), and on the right, the scatter plot of the random sample and the eigenvectors of the empirical covariance matrix, scaled by the two times the square roots of the corresponding eigenvalues. These square roots represent the standard deviations of the density in the eigendirections, so the lengths of the drawn vectors equal two standard deviations

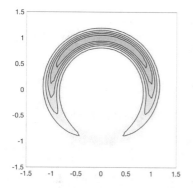 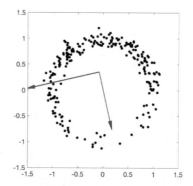

Fig. 4.2 The equiprobability curves of the original non-Gaussian distribution (left) and the sample drawn from it (right), together with the scaled eigenvectors of the empirical covariance matrix applied at the empirical mean

the empirical mean. While the mean and the covariance estimated from the sample are quite reasonable, a Gaussian model is clearly not in agreement with the sample. Indeed, if we would generate a new sample from a Gaussian density with the calculated mean and covariance, the scatter plot would be very different from the scatter plot of the sample. In this case, the mean and covariance of the sample provide a poor summary description of the sample.

4.2 Whitening and P–P Plots

In the previous examples, the parametric Gaussian model was clearly not a good choice for the second set of data. In one or two dimensions, it is relatively easy to check whether a Gaussian model (4.2), or any other parametric model that we choose to use, is reasonable for explaining the distribution of given data, simply by visually inspecting scatter plots. Although in higher dimensions we can look at the scatter plots of two-dimensional projections and try to assess if the normality assumption is reasonable[1], it is nonetheless preferable to develop more systematic methods to investigate the goodness of the Gaussian approximation. One possible way is to go through the cumulative distribution functions, comparing that of the theoretical model with the one estimated from the sample, as explained below. To simplify the discussion, we start by introducing the concept of *whitening* a random variable.

A standard normal Gaussian random variable $W \in \mathbb{R}^n$,

$$W \sim \mathcal{N}(0, I_n),$$

---

[1] As the legendary Yogi Berra summarized in one of his famous aphorisms, "you can observe a lot by just watching."

where I_n is the $n \times n$ unit matrix, is also referred to as *Gaussian white noise*. In this context, whitening is a transformation that maps an arbitrary Gaussian random variable to Gaussian white noise. Consider a random variable

$$X \sim \mathcal{N}(\mu, C),$$

where $\mu \in \mathbb{R}^n$, $C \in \mathbb{R}^{n \times n}$ is a symmetric positive definite matrix, and let

$$C = R^T R \tag{4.7}$$

be its Cholesky factorization. Given a Gaussian white noise W, define the random variable

$$Z = \mu + R^T W.$$

It is straightforward to verify that the mean and the covariance of this random variable are

$$E(Z) = \mu + R^T E(W) = \mu,$$

and

$$\begin{aligned} \text{cov}(Z) = E\big((Z - \mu)(Z - \mu)^T\big) &= E\big((R^T W)(R^T W)^T\big) \\ &= R^T E(W W^T) R = R^T R = C, \end{aligned}$$

respectively.

Moreover, using the change of variables formula introduced in the previous chapter, it can be shown that

$$Z \sim \mathcal{N}(\mu, C).$$

Therefore, X and Z can be identified as random variables, and we may write X in terms of Gaussian white noise,

$$X = \mu + R^T W,$$

or, by solving for W, we can express the Gaussian white noise in terms of X,

$$W = R^{-T}(X - \mu), \tag{4.8}$$

where $R^{-T} = (R^T)^{-1} = (R^{-1})^T$. Formula (4.8) defines a *whitening transformation*, or *Mahalanobis transformation*, of the random variable X into Gaussian white noise.

As an immediate application of the whitening process, consider now the question addressed earlier: Given a sample

$$S = \{x^1, x^2, \ldots, x^N\}, \quad x^j \in \mathbb{R}^n,$$

how close is it to a set of realizations from a Gaussian distribution? We perform the analysis in two dimensions, $n = 2$, to keep the computations tractable; the extension of the technique to $\mathbb{R}^n$ is fairly straightforward, requiring only little more than changing the notation. The general case is discussed in the Notes and Comments at the end of this chapter. To simplify the notation, let μ and C denote the empirical mean and covariance of the sample, and let (4.7) be the Cholesky decomposition of the latter. Apply the whitening transformation defined by R to the data, and consider the whitened sample

$$w^j = \mathsf{R}^{-T}(x^j - \mu), \quad 1 \le j \le N.$$

If the points x^j are realizations of a Gaussian random variable X with mean μ and covariance C, then the points w^j should be realizations of Gaussian white noise. It is rather easy to test if this is indeed the case.

For any $\delta > 0$, we can calculate the probability

$$\mathsf{P}\{\|W\| < \delta\} = \int_{D_\delta} \pi_W(w) dw, \tag{4.9}$$

where D_δ is the ball in $\mathbb{R}^2$ with radius δ. Formula (4.9) defines the cumulative distribution of the real-valued random variable $\|W\|$. If W is indeed Gaussian white noise, the value of the integral, computed by switching to polar coordinates (r, θ), is

$$\mathsf{P}\{\|W\| < \delta\} = \frac{1}{2\pi} \int_{D_\delta} e^{-\frac{1}{2}\|w\|^2} dw = \frac{1}{2\pi} \int_0^\delta \int_0^{2\pi} e^{-\frac{1}{2}r^2} r \, d\theta \, dr$$
$$= \int_0^\delta e^{-\frac{1}{2}r^2} r \, dr = 1 - e^{-\frac{1}{2}\delta^2}. \tag{4.10}$$

This number is equal to the proportion of the mass of the Gaussian probability density inside a disc of radius δ. Conversely, the radius δ of a disc centered at the origin containing the fraction p of the mass in its interior, $0 \le p \le 1$, can be found by solving

$$1 - e^{-\frac{1}{2}\delta^2} = p, \quad \text{or} \quad \delta = \delta(p) = \sqrt{2 \log\left(\frac{1}{1-p}\right)}.$$

Therefore, if the discrete sample is from an underlying Gaussian distribution, it is reasonable to expect approximately the same proportion of the whitened sample w^j to be inside the disc D_δ. Introduce the function

$$\nu(p) = \frac{1}{N} \#\{w^j \in D_{\delta(p)}\}. \tag{4.11}$$

Fig. 4.3 Plot of the sample-based quantiles $\nu(p)$ as a function of p. The solid curve corresponds to a sample arising from the Gaussian density, and the dashed curve to a non-Gaussian sample. Observe that in the non-Gaussian case, the sets $B_{\alpha(p)}$ with small p contain little or no sample points, and the slope is small

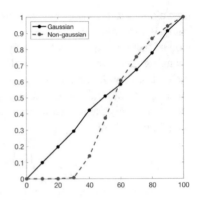

In the case when the points come from the Gaussian distribution, we should have $\nu(p) = p$, while large deviations of the values $\nu(p)$ from p suggest that the points of S are not likely to be realizations of a random variable whose probability density is normal. The plot of the curve $p \mapsto \nu(p)$ is a version of a multivariate *P-P plot*, or probability–probability plot, or percent–percent plot.

Figure 4.3 shows the results of this test with the two samples of Example 4.1. The solid curve is calculated by using the sample that indeed was generated by using the Gaussian model: as expected, the P-P plot is following well the diagonal $y = x$. The non-Gaussianity of the second sample, plotted as a dashed curve, is far from the diagonal: We observe that for small p, the function $\nu(p)$, instead of increasing with unit slope, remains close to zero, as there are no sample points near the mean, but after a while, the accumulation becomes faster, and the slope exceeds the unity, causing an overshoot of the curve.

4.3   Quadratures and Law of Large Numbers

So far, we have concentrated on characterizing a probability density underlying a discrete sample. In the following, we reverse the approach, and we seek to approximate a probability density by a discrete sample.

Consider first a random variable X taking on values in $\mathbb{R}$, and let π_X denote its probability density function. In order to calculate the expectation of a derived random variable $Y = f(X)$, where f is assumed to be a continuous and bounded function, we need to evaluate the integral

$$\mathrm{E}\big(f(X)\big) = \int_{\mathbb{R}} f(x)\pi_X(x)dx.$$

Assuming, further, that we know that the possible values of X are restricted to a finite interval which, for simplicity, we assume to be $[-1, 1]$, we may approximate the integral using classical quadrature rules such as Gauss quadratures. Recall that if we have a weight function $w(x)$ defined over the interval $[-1, 1]$, a Gauss quadrature rule comprises a set of nodes $x^j \in [-1, 1]$ and corresponding weights ω_j, $1 \leq j \leq m$, such that the integral can be approximated by a finite sum,

$$\int_{-1}^{1} f(x)w(x)dx \approx \sum_{j=1}^{m} \omega_j f(x^j), \tag{4.12}$$

and the approximation is exact for all polynomials p of degree at most $2m - 1$,

$$\int_{-1}^{1} p(x)w(x)dx = \sum_{j=1}^{m} \omega_j p(x^j), \quad \deg(p) \leq 2m - 1.$$

Different weight functions lead to different quadrature rules. For instance, the weight function $w(x) = 1$ leads to Gauss-Legendre rules, where the nodes are zeros of the Legendre polynomials, while the choice $w(x) = (1 - x^2)^{-1/2}$ leads to the Gauss-Chebyshev rule. Assuming for simplicity that $w(x) = 1$, we may therefore write

$$E\big(f(X)\big) \approx \sum_{j=1}^{m} \omega_j \pi_X(x^j) f(x^j) = \sum_{j=1}^{m} w_j f(x^j), \quad w_j = \omega_j \pi_X(x^j). \tag{4.13}$$

Observe that, by using the notion of point masses, we can write the approximation (4.13) in the following form:

$$\int_{-1}^{1} f(x)\pi_X(x)dx \approx \int_{-1}^{1} \sum_{j=1}^{m} w_j \delta_{x^j}(x) f(x)dx.$$

with the probability density function π_X replaced by a discrete point mass approximation,

$$\pi_X(x) \approx \sum_{j=1}^{m} w_j \delta_{x^j}(x).$$

The approximation is understood to hold in the sense that the discrete density integrates polynomials up to the order $2m - 1$ exactly. To avoid the use of point masses, this approximation can be written in terms of the cumulative distribution functions as follows. Denoting the cdf of X by

$$\Phi_X(s) = P\{X \leq s\} = \int_{-\infty}^{s} \pi_X(x)dx,$$

we have the piecewise constant approximation of Φ_X,

$$\Phi_X(s) \approx \sum_{j=1}^{m} w_j H(s - x^j),$$

where H is the Heaviside step function,

$$H(x) = \begin{cases} 1, & \text{if } x \geq 0, \\ 0, & \text{if } x < 0. \end{cases} \tag{4.14}$$

While Gauss quadratures and other classical quadrature rules are excellent tools for approximating integrals in low-dimensional spaces, the curse of dimensionality makes them less useful in higher dimensions. In fact, if the data are n-dimensional, $x^j \in \mathbb{R}^n$, a Gauss quadrature over a hypercube $[-1, 1]^n$ would require m^n grid points, a number that becomes quickly prohibitively large. What makes the matter even worse is the fact that at most of the grid points $\pi_X(x^j) \approx 0$, as illustrated schematically in Fig. 4.4. An additional difficulty arises from the fact that, in general, there is no systematic way to determine a priori a hyper-rectangle encompassing any significant part of the probability mass of the distribution.

The above considerations are a great motivation to look at random sampling from a different point of view. We start with a simple introductory example.

Example 4.2 Assume that on a rectangular sheet of paper with a known area A, an irregular shape Ω is drawn, and the problem is to estimate the area of that shape without any measuring tools such as a ruler or a compass. It is a rainy day, so we take the paper sheet out and let the raindrops fall on it, counting the drops falling inside the shape, as well as all drops falling on the paper sheet. We then estimate the area $|\Omega|$ by writing

$$\frac{|\Omega|}{A} \approx \frac{\text{\# raindrops inside } \Omega}{\text{\# all raindrops hitting the paper}}.$$

Fig. 4.4 A distribution in $[0, 1] \times [0, 1]$ covered by a regular grid of $21 \times 21 = 441$ nodes. The nodes at which the density takes on a value larger than 0.1% of its maximum are indicated by dots, representing about that 18% of the grid points, indicating that most of the function evaluations at grid points are numerically insignificant

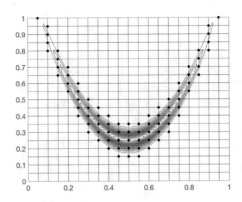

Here, we implicitly assume that the raindrops fall uniformly and independently. Thus, if π_X denotes the uniform distribution over the paper sheet,

$$\pi_X(x) = \frac{1}{A},$$

and $f(x)$ is the characteristic function of the shape Ω, we have

$$\frac{|\Omega|}{A} = \frac{1}{A} \int f(x)dx = \int f(x)\pi_X(x)dx,$$

and if N is the total number of raindrops on the paper, and the position of the jth raindrop is x^j, we have

$$\frac{\#\ \text{raindrops inside}\ \Omega}{\#\ \text{all raindrops hitting the paper}} = \frac{1}{N} \sum_{j=1}^{N} f(x^j).$$

The heuristics has therefore lead us to the stochastic quadrature formula,

$$\int f(x)\pi_X(x)dx \approx \frac{1}{N} \sum_{j=1}^{N} f(x^j),$$

a basic identity of Monte Carlo integration.

To demonstrate the algorithm, consider the domain Ω with a parametrized boundary curve

$$\begin{aligned} x(s) &= r\left(\cos s + \alpha(\cos 2s - 1)\right), \\ y(s) &= r\beta \sin s, \end{aligned} \qquad 0 \le s \le 2\pi,$$

where the parameters are chosen as $r = 0.5$, $\alpha = 0.65$, and $\beta = 1.5$. The domain is shown in Fig. 4.5. Using Green's theorem, we can compute the area of Ω exactly as a trigonometric integral, to get

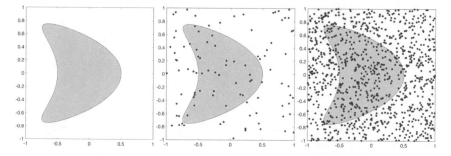

Fig. 4.5 The domain Ω inside the rectangle $[-1, 1] \times [-1, 1]$, and random points drawn from the uniform distribution. The number of the random points is 100 (center) and 1 000 (right)

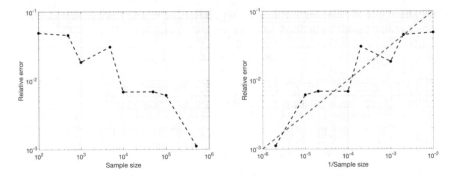

Fig. 4.6 Relative approximation error of the raindrop integrator as a function of the sample size N (left) as well as as a function of its inverse $1/N$ (right). The dashed red line in the latter indicates the power law $\sim 1/\sqrt{N}$

$$|\Omega| = \pi r^2 \beta \approx 1.178.$$

Let the paper sheet have the area $A = 4$ as in the figure, and simulate the raindrops by drawing points from the uniform distribution over the rectangle $[-1, 1] \times [-1, 1]$ as shown in the figure.

Figure 4.6 shows the relative error as a function of the sample size N (left), as well as a function of $1/N$ (right), plotted in logarithmic scales. The plots reveal that the relative error decreases roughly at the rate of $1/\sqrt{N}$, which is a very characteristic rate for stochastic integration methods.

We discuss the idea now in more general terms. Assume that we have an ensemble of points x^j independently drawn from a distribution described by a probability density π_X, that is, x^j is a realization of X^j, and that the random variables X^j are independent and identically distributed, $X^j \sim \pi_X$. The Law of Large Numbers guarantees that, for a continuous bounded function f,

$$\lim_{N \to \infty} \sum_{j=1}^N w_j^N f(x^j) = \int f(x) \pi_X(x) dx, \quad w_j^N = \frac{1}{N},$$

almost certainly. In other words,

$$\sum_{j=1}^N w_j^N \delta_{x^j} = \pi_X^N \to \pi_X$$

almost certainly *in the weak sense*[2], which means that, with probability one,

$$\int f(x) \pi_X^N(x) dx \to \int f(x) \pi_X(x) dx \quad \text{for every bounded continuous } f.$$

---

[2] Strictly speaking, we should say "in the weak* sense," for reasons explained in Notes and Comments at the end of the chapter.

Hence, we may view sampling from π_X as a way to generate a stochastic quadrature rule with uniform weights. In particular, as the points x^j are sampled from the probability density π_X, the method avoids wasting points in regions where the probability density π_X vanishes or is negligibly small. Moreover, the Central Limit Theorem gives us an idea of the convergence rate of the approximation. Since the random variables $f(X^j)$ are independent and identically distributed, we conclude that

$$\sqrt{N}\left(\int f(x)\pi_X^N(x)dx - \int f(x)\pi_X(x)dx\right) \xrightarrow{d} \mathcal{N}(0, \sigma^2),$$

where σ^2 is the variance of the random variable $f(X)$ and the convergence takes place "in distribution" as defined in Sec. 3.3. Thus, the theorem establishes a convergence rate of $1/\sqrt{N}$.

Integration based on random sampling is called *Monte Carlo integration*. Although it looks very attractive, it has indeed its share of difficulties; the proverbial free lunch does not exist here, either. One of the problems is how to draw the sample. We start examining this problem in this section, and return to it later with more sophisticated tools at our disposal.

There are two elementary distributions that we take for granted in the sequel. We assume that we have access to

- a random number generator drawing from the uniform distribution over the interval $[0, 1]$,

$$\pi_X(x) = \chi_{[0,1]}(x),$$

- a random number generator drawing from the standard normal distribution,

$$\pi_X(x) = \frac{1}{\sqrt{2\pi}}e^{-x^2/2}.$$

These (pseudo)random generators in Matlab are called with the commands `rand` and `randn`, respectively.

4.4 Drawing from Discrete Densities

Consider a random variable with a finite set of possible values $\{v_1, v_2, \ldots, v_n\}$, having therefore a probability density comprising point masses,

$$\pi_X(x) = \sum_{j=1}^{n} p_j \delta_{v_j}(x),$$

where the weights p_j must satisfy

$$p_j \geq 0, \quad \sum_{j=1}^{n} p_j = 1.$$

We start by dividing the unit interval $[0, 1]$ into n disjoint intervals I_j such that

$$\text{length}(I_j) = p_j, \quad 1 \leq j \leq n.$$

Consider a random variable uniformly distributed in $[0, 1]$, $W \sim \text{Uniform}$ $([0, 1])$. By definition of the uniform distribution, we conclude that

$$P\{W \in I_j\} = \text{length}(I_j) = p_j,$$

which suggests a simple algorithm for drawing randomly realizations of X:

1. Draw randomly the value w of W,
2. Find j such that $w \in I_j$, and set $x = v_j$.

The implementation can be done effectively in terms of the cumulative distribution function. Here is how to do it in Matlab:

```
Phi = 0;
j   = 0;
w   = rand;
while Phi < w
    j   = j+1;
    Phi = Phi + p(j);
end
```

In the algorithm, the variable Phi represents the cumulative distribution function. The exit value of j is the output of the algorithm, as visualized in Fig. 4.7.

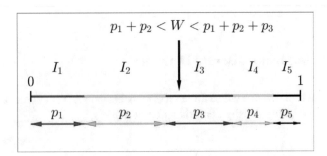

Fig. 4.7 Drawing from a discrete distribution: Divide the interval $[0, 1]$ into subintervals I_j of length p_j, then draw W from uniform distribution and check in which interval the value falls, giving the discrete random variable

If n is large, the above algorithm can be sped up by sorting first the indices so that the realizations with the highest probabilities appear first, implying that the cumulative distribution reaches the random number w in fewest possible steps. This algorithm can be used to draw a sample from discrete random variables with an infinite set of possible values, such as the Poisson distribution.

4.5 Sampling from a One-Dimensional Continuous Density

We are now ready to generalize the idea of sampling from a discrete density to the case of a random variable with a continuum of possible values. Hence, let π_X denote the probability density of a random variable X defined on $\mathbb{R}$. We start by recalling the definition of the cumulative distribution function of X,

$$\Phi_X(s) = P\{X \leq s\} = \int_{-\infty}^{s} \pi_X(x)dx.$$

The function Φ_X is non-decreasing since $\pi_X \geq 0$, and $0 \leq \Phi_X \leq 1$. To simplify the discussion, we assume that π_X is supported in some interval $I \subset \mathbb{R}$, and $\pi_X(x) > 0$ except possibly at some isolated points in I. With these assumptions, the cumulative distribution function is strictly increasing over I, and it attains all values in the open interval $(0, 1)$.

Let us define a new random variable,

$$T = \Phi_X(X). \tag{4.15}$$

We claim that $T \sim \text{Uniform}([0, 1])$. In general, a random variable Y follows the uniform distribution over an interval $[a, b]$ if

$$P\{Y < s\} = \frac{s - a}{b - a}, \quad a \leq s \leq b.$$

To prove that T is uniformly distributed, observe first that, due to the monotonicity of Φ_X,

$$P\{T \leq s\} = P\{\Phi_X(X) \leq s\} = P\{X \leq \Phi_X^{-1}(s)\}, \quad 0 < s < 1.$$

On the other hand, by the definition of probability density,

$$P\{X \leq \Phi^{-1}(s)\} = \int_{-\infty}^{\Phi_X^{-1}(s)} \pi_X(x)dx = \int_{-\infty}^{\Phi_X^{-1}(s)} \Phi_X'(x)dx.$$

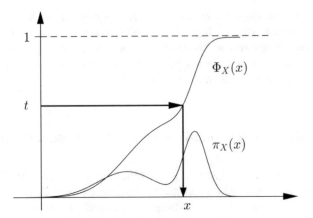

Fig. 4.8 A random draw of x from the distribution $\pi_X(x)$, corresponding to the cumulative distribution $\Phi_X(x)$. Here, $t \sim \text{Uniform}([0, 1])$

After the change of variables

$$t = \Phi_X(x), \quad dt = \Phi_X'(x)dx,$$

we arrive at

$$P\{T \le s\} = \int_{-\infty}^{\Phi_X^{-1}(s)} \Phi_X'(x)dx = \int_0^s dt = s,$$

which is equivalent to saying that T is a random variable with uniform distribution over $[0, 1]$, thus proving the claim.

We are now ready to present an algorithm to draw from the distribution π_X, graphically illustrated in Fig. 4.8, comprising two simple steps:

1. Draw $t \sim \text{Uniform}([0, 1])$,
2. Calculate $x = \Phi_X^{-1}(t)$.

This algorithm is sometimes referred to as the *golden rule*, or *inverse cumulative distribution rule*. In spite of the simplicity of the idea, its implementation can present some problems, as the following example shows.

Example 4.3 Consider a one-dimensional normal distribution with a bound constraint,

$$\pi_X(x) \propto H(x - c)e^{-x^2/2},$$

where "$\propto$" means "proportional to" or equal to up to a normalizing constant, and H is the Heaviside step function (4.14). The problem is to generate a sample drawing independently from π_X.

The cumulative distribution function is

$$\Phi_X(x) = C \int_c^x e^{-y^2/2}dy, \quad x \ge c,$$

and $\Phi_X(x) = 0$ for $x < c$, where $C > 0$ is the normalizing constant

$$C = \left(\int_c^\infty e^{-y^2/2} dy \right)^{-1}.$$

The function Φ_X has to be calculated numerically. Fortunately, there is reliable software available to do the calculation. In Matlab, the built-in *error function*, `erf`, is defined as

$$\mathrm{erf}(s) = \frac{2}{\sqrt{\pi}} \int_0^s e^{-z^2} dz.$$

We observe that

$$\Phi_X(x) = C \left(\int_0^x - \int_0^c \right) e^{-y^2/2} dy = \sqrt{2} C \left(\int_0^{x/\sqrt{2}} - \int_0^{c/\sqrt{2}} \right) e^{-z^2} dz$$

$$= \sqrt{\frac{\pi}{2}} C \left(\mathrm{erf}(\frac{x}{\sqrt{2}}) - \alpha \right), \quad \alpha = \mathrm{erf}(\frac{c}{\sqrt{2}}).$$

On the other hand, since

$$C = \left(\sqrt{\frac{\pi}{2}} (1 - \alpha) \right)^{-1},$$

we have

$$\Phi_X(x) = \frac{\mathrm{erf}(x/\sqrt{2}) - \alpha}{1 - \alpha}.$$

What about the inverse? Writing

$$\Phi_X(x) = t,$$

after some algebraic manipulations, we find that

$$\mathrm{erf}(\frac{x}{\sqrt{2}}) = t(1 - \alpha) + \alpha.$$

The double luck is that there are effective algorithms to calculate the inverse of the error function, for example `inverf` in Matlab:

$$x = \Phi_X^{-1}(t) = \sqrt{2} \, \mathrm{inverf}\big(t(1 - \alpha) + \alpha\big).$$

Hence, random generation in Matlab is very simple:

```
alpha = erf(c/sqrt(2));
t     = rand;
x     = sqrt(2)*erfinv(t*(1-alpha)+alpha);
```

Here, a warning is in order: if the bound c is large, the above algorithm may fail, because the error function saturates quickly to unity: When c is large, numerically $\alpha = 1$, and the above algorithm gives numerical noise as a value of x. To understand what "large" means here, $1 - \mathrm{erf}(3) \approx 2.2 \times 10^{-5}$, and $1 - \mathrm{erf}(4) \approx 1.5 \times 10^{-8}$. Therefore, a robust implementation must address the special case of c large. We do not discuss the modification that is needed.

4.6 Sampling from Gaussian Distributions

We turn now to the problem of generating samples from a general Gaussian probability density. Recall that if X is a multivariate Gaussian random variable, $X \sim \mathcal{N}(\mu, \mathsf{C})$, where $\mu \in \mathbb{R}^n$ and $\mathsf{C} \in \mathbb{R}^{n \times n}$ is a symmetric positive definite matrix, the probability density function is

$$\pi_X(x) = \left(\frac{1}{(2\pi)^n \det(\mathsf{C})} \right)^{1/2} \exp \left(-\frac{1}{2}(x - \mu)^\mathsf{T} \mathsf{C}^{-1}(x - \mu) \right).$$

The main tool for generating a sample from π_X is the whitening process, or Mahalanobis transformation, introduced in Sect. 4.2. We compute the Cholesky factorization of the covariance matrix,

$$\mathsf{C} = \mathsf{R}^\mathsf{T} \mathsf{R}, \tag{4.16}$$

and represent X in terms of a standard normal random variable W as

$$X = \mu + \mathsf{R}^\mathsf{T} W, \quad W \sim \mathcal{N}(0, \mathsf{I}_n).$$

With this representation, random draws from π_X can be done simply as follows:

1. Draw a realization $w \in \mathbb{R}^n$ from the Gaussian standard normal density $\mathcal{N}(0, \mathsf{I}_n)$,
2. Calculate the realization x by the formula

$$x = \mu + \mathsf{R}^\mathsf{T} w.$$

Notice that the decomposition (4.16) need not necessarily be the Cholesky factorization, but, in fact, any symmetric factorization of C will do. This observation is particularly useful for Gaussian densities that are given in terms of the *precision matrix*, which, by definition, is the inverse of the covariance matrix, $\mathsf{P} = \mathsf{C}^{-1}$ rather than C itself. In those cases, it is natural to compute the Cholesky (or any other symmetric) factorization of P,

$$\mathsf{P} = \mathsf{L}^\mathsf{T} \mathsf{L}.$$

This factorization implicitly defines a symmetric factorization of the covariance matrix,

$$C = P^{-1} = L^{-1}L^{-T},$$

and we may express the factor R in terms of L as $R = L^{-T}$. The above algorithm for drawing samples x can be modified as follows:

1. Draw a realization $w \in \mathbb{R}^n$ from the Gaussian standard normal density $\mathcal{N}(0, I_n)$,
2. Solve the linear system

$$w = Lz$$

for z, and define

$$x = \mu + z.$$

The application of this alternative approach will be illustrated later.

4.7 Some Useful Sampling Algorithms

If the probability density of interest is not Gaussian, or one-dimensional, the methods discussed so far cannot be employed, and the sampling may become more challenging. In this section, we review some methods that are of interest in those cases, and that turn out to be particularly useful building blocks for the sequential Bayesian methods that will be introduced later in the book.

4.7.1 Importance Sampling

Assume that two probability distributions π_1 and π_2 in $\mathbb{R}^n$ are given, and they are related to each other through the formula:

$$\pi_2(x) = Cf(x)\pi_1(x), \tag{4.17}$$

where $f(x) \geq 0$, and C is a normalizing constant guaranteeing that

$$\int \pi_2(x)dx = C \int f(x)\pi_1(x)dx = 1.$$

We address the following problem: Assuming that a sampling algorithm from the distribution π_1 is available, how can we use it to produce a sample from the distribution π_2? To that end, let φ be a test function, that we assume to be continuous and bounded, and consider the integral

$$\int \varphi(x)\pi_2(x)dx = C \int \varphi(x)f(x)\pi_1(x)dx, \tag{4.18}$$

that we want to approximate by a finite sum. By assumption, we have a method of sampling from the distribution π_1. Let $\{(x^j, w_j)\}_{j=1}^N$ be a sample from π_1. Then

$$\int_{\mathbb{R}} \varphi(x)\pi_1(x)dx \approx \sum_{j=1}^N w_j\varphi(x^j),$$

or, in terms of densities,

$$\pi_1(x) \approx \sum_{j=1}^N w_j\delta_{x^j}(x)$$

in the weak sense. This implies that we may approximate the integral (4.18) with

$$\int \varphi(x)\pi_2(x)dx = C \int \varphi(x)f(x)\,\pi_1(x)dx \approx C \sum_{j=1}^N w_j f(x^j)\varphi(x^j),$$

suggesting that

$$\pi_2(x) \approx \sum_{j=1}^N \left(Cw_j f(x^j)\right)\delta_{x^j}(x) = \sum_{j=1}^N \widetilde{w}_j\delta_{x^j}(x), \quad \widetilde{w}_j = Cw_j f(x^j),$$

that is, the desired sample W from π_2 consists of the points in the sample from π_1, and only the weights need to be updated. Observe that although the normalizing constant C may not be known, we need the new weights $\widetilde{w}_j$ to define a probability measure, so it is natural to assume that

$$\sum_{j=1}^N \widetilde{w}_j = 1.$$

This leads to the following algorithm.

Importance sampling: Assume that the probability densities π_1 and π_2 satisfy (4.17), and that a sample $\{(x^j, w_j)\}_{j=1}^N$ from π_1 is available. Then the sample $\{(x^j, \widetilde{w}_j)\}_{j=1}^N$, where the x^js are the same and the weights $\widetilde{w}_j$ are computed from the w^j by

1. calculating $\widetilde{w}_j = w_j f(x^j)$,
2. normalizing

$$\widetilde{w}_j \to \frac{\widetilde{w}_j}{\sum_{k=1}^N \widetilde{w}_k},$$

approximates π_2.

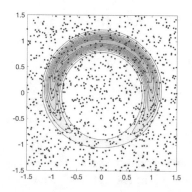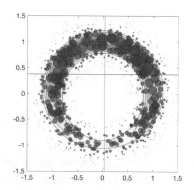

Fig. 4.9 Left: Uniformly distributed sample with equal weights over the square $[-1.5, 1.5] \times [-1.5, 1.5]$, and the density π_2. Right: The sample obtained by importance sampling, the areas of the dots being proportional to the corresponding weights. The crosshair indicates the estimated mean of the density π_2

In other words, if two distributions are related via a known functional form (4.17), and one of them is approximated by a sample of ordered pairs of points and weights, to approximate the other it suffices to update the weights. This algorithm will turn out to be very useful in the context of sequential updating.

Example 4.4 Consider the horseshoe distribution of Example 4.1, denoted here by π_2. We choose the distribution π_1 to be a uniform distribution over the square $[-1, 5, 1.5] \times [-1.5, 1.5]$, and draw a sample of $N = 1\,000$ points from it, shown on the left of Fig. 4.9. Using the importance sampling, we then adjust the weights $w_j = 1/N$. The sample points are shown on the right of Fig. 4.9, where the area of each dot is proportional to the weight $\widetilde{w}_j$. The figure also shows the estimated mean of π_2, calculated by using the sample,

$$\overline{x} = \sum_{j=1}^{N} \widetilde{w}_j x^j.$$

Observe that while the algorithm is rather simple and easy to implement, it is of limited use for high-dimensional integration, since it is open to the same criticism as deterministic quadrature rules: Most of the evaluations to compute $\widetilde{w}_j$ are giving a negligible contribution, and the method requires a knowledge of a good bounding box for generating the initial sample.

4.7.2 Drawing from Mixtures: SIR and Weighted Bootstrap

Another way to approximate a probability density is in terms of a mixture of some elementary probability densities π_ℓ,

$$\pi_X(x) = \sum_{\ell=1}^{M} \alpha_\ell \pi_\ell(x), \tag{4.19}$$

where the weights α_ℓ are non-negative. The densities π_ℓ are chosen so that drawing from them is easy. If the densities π_ℓ are scaled to have the integral equal to one, we have

$$\int \pi_X(x)dx = \sum_{\ell=1}^{M} \alpha_\ell \int \pi_\ell(x)dx = \sum_{\ell=1}^{M} \alpha_\ell,$$

and it is natural to assume that the weights α_ℓ are normalized so that

$$\sum_{\ell=1}^{M} \alpha_\ell = 1.$$

In the commonly used *Gaussian mixtures*, each π_ℓ is a Gaussian density,

$$\pi_\ell(x) = \mathcal{N}(x \mid \mu_\ell, C_\ell),$$

with mean μ_ℓ and covariance matrix C_ℓ.

In order to design a sampling algorithm from a mixture, we introduce the new integer-valued random variable L that can take on the values $\{1, 2, \ldots, M\}$, and postulate that

$$\pi_L(\ell) = P\{L = \ell\} = \alpha_\ell, \quad 1 \le \ell \le M.$$

Furthermore, we define the conditional densities

$$\pi_{X|L}(x \mid \ell) = \pi_\ell(x),$$

so that the mixture model (4.19) becomes simply a marginal density,

$$\pi_X(x) = \sum_{\ell=1}^{M} \pi_{X|L}(x \mid \ell)\pi_L(\ell).$$

The sampling strategy is to first draw pairs (x^j, ℓ^j) from the joint probability density $\pi_{X|L}(x \mid \ell)\pi_L(\ell)$, then marginalize by projecting the sample onto the first component.

In summary, we draw pairs (x, ℓ) from the joint density by repeating the following steps:

1. Draw the realizations ℓ^j of the random variable L from the discrete distribution with point masses $\pi_K(\ell) = \alpha_\ell$;
2. For each selected ℓ^j, draw x^j from the conditional density $\pi_{X|L}(x \mid \ell^j) = \pi_{\ell^j}(x)$.

Finally, we marginalize by discarding the indices ℓ^j, keeping only the sample $\{x^1, x^2, \ldots, x^M\}$.

Consider the special case when the Gaussian densities are defined as

$$\pi_\ell(x) = \mathcal{N}(x \mid \mu_\ell, h^2 \mathsf{I}_n), \quad h > 0.$$

In the limit $h \to 0+$, when the distributions in the mixture become point masses,

$$\pi_{X|L}(x \mid \ell) = \delta_{\mu^\ell}(x),$$

we have a particular instance of the weighted bootstrap algorithm, and the probability density is approximated as

$$\pi_X(x) \approx \sum_{\ell=1}^{M} \alpha_\ell \delta_{\mu^\ell}(x). \tag{4.20}$$

One might wonder what the reason is for drawing samples from such an approximation, when the approximation itself is already a discrete approximation, and the sample $S = \{(\mu^1, \alpha_1), \ldots, (\mu^M, \alpha_M)\}$ is readily available. As will become evident later, sometimes it is preferable to work with a sample where all sample points have the same weights. Another instance when it is necessary to draw from an approximation is when one needs a sample of size different from M. In this case, drawing from $\pi_{X|L}(x \mid \ell^j) = \delta_{\mu^{\ell^j}}(x)$ can produce only one possible outcome, namely $x^j = \mu^{\ell^j}$. Observe that if the weights α_ℓ are of different magnitudes, the sampling is likely to produce several copies of the same points with high weights, while some other points with low weights are likely not to be represented at all. Thus, the importance of points x^j in the new sample is encoded in the possible repetition of particles having originally a larger weight, called *importance weight*. For this reason, the resulting algorithm is known as *sampling-importance resampling (SIR)*.

In the discussion above, we assumed that a mixture model (4.19) was given, and outlined a two-phase algorithm for drawing from it. Conversely, assume that a sample $\{(\xi^1, w_1), \ldots, (\xi^M, w_M)\}$ from the density π_X is given, and we want to find a mixture model in order to draw new points from π_X. If we use the approximation

$$\pi_X(x) \approx \sum_{\ell=1}^{M} w_\ell \delta_{\xi^\ell}(x), \tag{4.21}$$

the SIR algorithm has the tendency to produce repeated examples of the same point, often referred to as *data thinning* or *sample impoverishment*: A sample consisting of mostly few repeated entries reflects poorly the variability of the underlying distribution. To avoid this problem, we build a Gaussian mixture model using the known sample.

To generate a Gaussian mixture, we first estimate the mean and covariance of the underlying density π_X using the base sample,

$$\mu = \sum_{j=1}^{M} w_j \xi^j, \quad C = \sum_{j=1}^{M} w_j (\xi^j - \mu)(\xi^j - \mu)^\mathsf{T}. \tag{4.22}$$

Next, we propose a Gaussian mixture model by defining a small Gaussian density around each base point, writing

$$\pi_X(x) \approx \widehat{\pi}_X(x) = \sum_{j=1}^{M} w_j \mathcal{N}(x \mid \xi^j, h^2 C), \tag{4.23}$$

where $h^2 > 0$ is a small parameter controlling the width of the Gaussians around the base points. With this approximation, the weighted bootstrap produces a sample similar to the one obtained with the SIR, except for the fact that the base points are not repeated, but replaced by draws from small Gaussians centered at the base points. Letting $h^2 \to 0$, the algorithm approaches SIR.

While, at first sight, all looks under control, it turns out that the algorithm, as proposed, has a small problem. Ideally, the mean and the covariance of the approximation (4.23) should be close to μ and C. This is indeed the case for the sample mean,

$$\int x \widehat{\pi}_X(x) dx = \sum_{j=1}^{M} w_j \int x \mathcal{N}(x \mid \xi^j, h^2 C) dx$$

$$= \sum_{j=1}^{M} w_j \xi^j = \mu.$$

Consider the covariance, writing it as

$$\int (x - \mu)(x - \mu)^\mathsf{T} \widehat{\pi}_X(x) dx = \sum_{j=1}^{M} w_j \int (x - \mu)(x - \mu)^\mathsf{T} \mathcal{N}(x \mid \xi^j, h^2 C) dx,$$

$$\tag{4.24}$$

and notice that each integral in the sum is

$$\int (x - \mu)(x - \mu)^{\mathsf{T}} \mathcal{N}(x \mid \xi^j, h^2 C) dx$$

$$= \int \left(x - \xi^j + (\xi^j - \mu) \right) \left(x - \xi^j + (\xi^j - \mu) \right)^{\mathsf{T}} \mathcal{N}(x \mid \xi^j, h^2 C) dx$$

$$= \int (x - \xi^j)(x - \xi^j)^{\mathsf{T}} \mathcal{N}(x \mid \xi^j, h^2 C) dx$$

$$+ \int (x - \xi^j)(\xi^j - \mu)^{\mathsf{T}} \mathcal{N}(x \mid \xi^j, h^2 C) dx$$

$$+ \int (\xi^j - \mu)(x - \xi^j)^{\mathsf{T}} \mathcal{N}(x \mid \xi^j, h^2 C) dx$$

$$+ \int (\xi^j - \mu)(\xi^j - \mu)^{\mathsf{T}} \mathcal{N}(x \mid \xi^j, h^2 C) dx$$

$$= h^2 C + (\xi^j - \mu)(\xi^j - \mu)^{\mathsf{T}}, \tag{4.25}$$

since the second and third terms in (4.25) vanish. Substituting this expression into (4.24) yields

$$\int (x - \mu)(x - \mu)^{\mathsf{T}} \widehat{\pi}_X(x) dx = h^2 C + \sum_{j=1}^{M} w_j (\xi^j - \mu)(\xi^j - \mu)^{\mathsf{T}}$$

$$= (1 + h^2) C,$$

Thus, we conclude that replacing the point masses by Gaussians automatically increases the covariance by a factor of $1 + h^2$. This can be corrected by shifting the base points ξ^j slightly towards their common mean μ. To accomplish this, we define

$$\bar{\xi}^j = a\xi^j + (1 - a)\mu, \quad a = \sqrt{1 - h^2}, \tag{4.26}$$

and redefine the Gaussian mixture density as

$$\widehat{\pi}_X(x) = \sum_{j=1}^{M} w_j \mathcal{N}(x \mid \bar{\xi}^j, h^2 C).$$

A similar computation shows that

$$\int x \widehat{\pi}_X(x) dx = \sum_{j=1}^{M} w_j \bar{\xi}_j = a\mu + (1 - a)\mu = \mu,$$

and since

$$\bar{\xi}^j - \mu = a(\xi^j - \mu),$$

it follows that

$$\int (x-\mu)(x-\mu)^\mathsf{T} \widehat{\pi}_X(x)dx = \sum_{j=1}^M w_j \left(h^2 \mathsf{C} + (\bar{\xi}^j - \mu)(\bar{\xi}^j - \mu)^\mathsf{T} \right)$$

$$= h^2 \mathsf{C} + a^2 \sum_{j=1}^M w_j (\xi^j - \mu)(\xi^j - \mu)^\mathsf{T}$$

$$= \mathsf{C},$$

as desired.

We are now ready to summarize the procedure in the form of an algorithm.

Weighted bootstrap algorithm: To generate a sample $\{x^1, \ldots, x^N\}$ based on the approximation (4.21),

1. Compute the empirical mean and covariance μ and C of the distribution from the base sample using formulas (4.22).
2. Choose $h > 0$, and compute the shifted base sample $\{\bar{\xi}^1, \ldots, \bar{\xi}^M\}$ according to formula (4.26).
3. For $j = 1, 2, \ldots, N$ repeat

   - Draw an index ℓ^j using the relative weights $w_1, \ldots, w_M$ as probabilities,
   - Draw x^j from the Gaussian distribution $\mathcal{N}(\bar{\xi}^{\ell^j}, h^2 \mathsf{C})$.

   Figure 4.10 shows a graphical rendition of mixture model.

4.8   Rejection Sampling: Prelude to Metropolis–Hastings

A common strategy to produce samples from a given distribution is to make proposals using a surrogate distribution, from which it is easy to draw samples, then either reject or accept the proposal. We illustrate this with a simple example.

Example 4.5 Consider the Gaussian density with a bound constraint introduced in Example 4.3,

$$\pi_X(x) = \begin{cases} Ce^{-x^2/2}, & \text{when } x > c, \\ 0 & \text{otherwise,} \end{cases}$$

where $C > 0$ is a normalizing constant. The most straightforward and intuitive approach for generating a sample from this distribution is the following trial-and-error algorithm:

1. Draw x from the normal distribution $\mathcal{N}(0, 1)$,
2. If $x > c$, accept, otherwise reject.

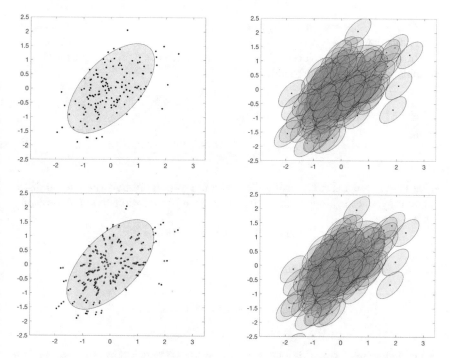

Fig. 4.10 Top left: A small sample drawn from a Gaussian, the ellipse of two standard deviations shown in the figure. Top right: Ellipses of two standard deviations of the distributions $\mathcal{N}(x \mid \xi^j, h^2 C)$ drawn around the points ξ^j, with $h = 0.3$. Bottom left: The shifted points $\bar{\xi}^j$ are plotted by red dots. Bottom right: Ellipses of two standard deviations of the distributions $\mathcal{N}(x \mid \bar{\xi}^j, h^2 C)$ drawn around the points $\bar{\xi}^j$

Observe that this is a version of the importance sampling algorithm, as the distribution can be written as

$$\pi_X(x) = C H(x - c)\mathcal{N}(x \mid 0, 1),$$

where H is the Heaviside step function, and the importance weights of points drawn from the normal distribution are either zero or one, depending on which side of the lower bound c the point falls.

While the algorithm works, it can be extremely slow. To understand why, consider the *acceptance rate* for different values of c. If $c = 0$, in the average every other proposal will be accepted, since the underlying normal distribution is symmetric. If $c < 0$, the acceptance rate will be higher than 50%. On the other hand, if $c > 0$, it will be smaller. In general, the probability of hitting the domain of acceptance $\{x > c\}$ is

$$I(c) = \frac{1}{\sqrt{2\pi}} \int_c^\infty e^{-x^2/2} dx,$$

and $I(c) \to 0$ superexponentially as $c \to \infty$. For instance, if $c = 3$, the acceptance rate is less than 0.1%.

The situation degrades dramatically for multidimensional problems. Consider, for example, a multivariate distribution in $\mathbb{R}^n$,

$$\pi(x) \propto \pi_+(x)e^{-\|x\|^2/2}, \quad x \in \mathbb{R}^n,$$

where
$$\pi_+(x) = \begin{cases} 1, & \text{if } x_j > 0 \text{ for all } j, \\ 0, & \text{else.} \end{cases}$$

What is the acceptance rate using the trial-and-error algorithm? The normal distribution is symmetric around the origin, so any sign combination is equally probable. Since
$$P\{x_j > 0\} = \frac{1}{2}, \quad 1 \le j \le n,$$

the rate of successful proposals is

$$P\{x_1 > 0, x_2 > 0, \ldots, x_n > 0\} = \prod_{j=1}^n P\{x_j > 0\} = \left(\frac{1}{2}\right)^n,$$

and it is clear that even a simple positivity constraint puts us out of business if n is large. The central issue therefore becomes how to modify the initial random draws, step 1 above, to improve the acceptance rate.

Inspired by the trial-and-error drawing strategy of the previous example while acknowledging its limitations, we propose an elementary sampling strategy that works—at least in theory—also for multidimensional distributions. The proposed algorithm is known as the *rejection sampling algorithm*.

Assume that the true probability density $\pi_X(x)$, from which we want to draw a sample, is known up to a multiplicative constant, i.e., we have the function

$$p(x) = C\pi_X(x),$$

with $C > 0$ unknown. In principle, if we had a way to integrate p over $\mathbb{R}^n$, the constant C could be determined, however, computing the integral of p may be hard, and in fact, one aim of the whole sampling is to give a way to estimate the integral. Assume that we have a *proposal distribution*, $q(x)$ that is easy to use for random draws—for instance, a Gaussian distribution. Furthermore, assume that the proposal distribution satisfies the condition

$$p(x) \le Mq(x) \quad \text{for some } M > 0.$$

The following algorithm is a prototype for the Metropolis–Hastings algorithm that will be introduced later:

1. Draw x from the proposal distribution $q(x)$,
2. Calculate the *acceptance ratio*

$$\alpha = \frac{p(x)}{Mq(x)}, \quad 0 < \alpha \le 1.$$

3. Flip the α-*coin*: Draw $t \sim \text{Uniform}([0, 1])$, and accept x if $\alpha > t$, otherwise reject.

The last step says simply that x is accepted with probability α, rejected with probability $1 - \alpha$, thus the name α-coin.

Why does the algorithm work? What we need to show is that the distribution of the accepted proposals x is π_X. To this end, we introduce the binary random variable A,

$$A = \begin{cases} 1, & \text{if a draw from } q \text{ is accepted,} \\ 0, & \text{if a draw from } q \text{ is rejected.} \end{cases}$$

The distribution of the accepted sample points *drawn from q*, i.e., the distribution of the sample that we have just generated, is

$$\pi_{X|A}(x \mid 1) = \text{distribution of } x\text{'s, given that they are accepted.}$$

To calculate this distribution, we use Bayes' formula,

$$\pi_{X|A}(x \mid 1) = \frac{\pi_{A|X}(1 \mid x)q(x)}{\pi_A(1)}, \tag{4.27}$$

setting the prior distribution of x equal to q, since we draw x from q. One interpretation of (4.27) is that before testing for acceptance, we believe that q is the correct distribution of x, hence our prior. What are the other densities appearing above?

The probability of acceptance *provided that x is given*, is clearly the acceptance ratio,

$$\pi_{A|X}(1 \mid x) = \alpha = \frac{p(x)}{Mq(x)}.$$

The marginal density of the acceptance is, in turn,

$$\pi_A(1) = \int \pi_{X,A}(x, 1)dx = \int \pi_{A|X}(1 \mid x)q(x)\,dx$$

$$= \int \frac{p(x)}{Mq(x)}q(x)\,dx = \frac{C}{M}\int \pi_X(x)\,dx$$

$$= \frac{C}{M}.$$

Substituting the derived expressions into (4.27), we have

$$\pi_{X|A}(x \mid 1) = \frac{\frac{p(x)}{Mq(x)}q(x)}{\frac{C}{M}} = \frac{p(x)}{C} = \pi_X(x).$$

exactly as we wanted.

We conclude this section verifying that the example of Gaussian distribution with a bound constraint is a version of a rejection sampling algorithm. In that case, the true distribution, up to a normalizing constant, is

$$p(x) = H(x - c)e^{-x^2/2},$$

the proposal distribution is Gaussian,

$$q(x) = \frac{1}{\sqrt{2\pi}}e^{-x^2/2},$$

and the scaling constant M can be chosen as $M = \sqrt{2\pi}$ so that

$$Mq(x) = e^{-x^2/2} \geq p(x).$$

The acceptance rate is then

$$\alpha = \frac{p(x)}{Mq(x)} = H(x - c) = \begin{cases} 1, & \text{if } x > c, \\ 0, & \text{if } x \leq c. \end{cases}$$

Flipping the α-coin in this case says that we accept the proposal automatically if $x > c$ ($\alpha = 1$), and reject otherwise ($\alpha = 0$), which is exactly what our heuristic algorithm was doing.

Notes and Comments

The factor $1/(N - 1)$ in the unbiased estimator $\widehat{C}_u$, formula (4.5) follows from the consideration that the estimated mean value itself is a realization of a random variable $N^{-1}(X^{(1)} + \ldots + X^{(N)})$, where the variables $X^{(j)}$ are independent and identically distributed, see, e.g, [1].

The P–P plots in this chapter were discussed only in $\mathbb{R}^2$ in order to keep the integrals simple. The modification for $\mathbb{R}^n$ is straightforward, requiring only that we replace (4.10) by the integrals

$$P\{\|W\| < \delta\} = \left(\frac{1}{2\pi}\right)^{n/2} \int_{D\delta} e^{-\frac{1}{2}\|w\|^2} dw$$

$$= \left(\frac{1}{2\pi}\right)^{n/2} \int_0^\delta \int_{\mathbb{S}^{n-1}} e^{-\frac{1}{2}t^2} t^{n-1} dS dt$$

$$= \frac{|\mathbb{S}^{n-1}|}{(2\pi)^{n/2}} \int_0^\delta e^{-\frac{1}{2}t^2} t^{n-1} dt,$$

where $\mathbb{S}^{n-1}$ is the unit sphere in $\mathbb{R}^n$ and $|\mathbb{S}^{n-1}|$ is its surface area. By a change of variables, the integral can be expressed in terms of the lower incomplete gamma function,

$$\gamma(m, s) = \int_0^s e^{-t} t^{m-1} dt.$$

Recalling that the surface area of the n-sphere is

$$|\mathbb{S}^{n-1}| = \frac{2\pi^{n/2}}{\Gamma(n/2)},$$

where Γ is the complete gamma function, we arrive at the formula

$$P\{\|W\| < \delta\} = \frac{\gamma(n/2, \delta^2/2)}{\Gamma(n/2)},$$

which approaches 1 as δ approaches infinity.

Estimates for the convergence rate of Gauss quadratures can be found in standard books of numerical analysis, see. e.g., [72].

It was pointed out in a footnote that the convergence of the linear combinations of point masses towards the distribution should be understood in the weak* sense. The reason for this nomenclature is that the dual space, $C(I)^*$, of bounded continuous functions over the interval I, consists, by definition, of linear functionals $C(I) \to \mathbb{R}$, and by the Riesz–Markov–Kakutani representation theorem, each functional has a unique representation in terms of Radon measures over I [68]. Hence, the described weak convergence takes place in the dual space, thus the notation.

The Gaussian mixture approximation using the shifted midpoints is discussed by Mike West in [82].

Chapter 5
The Praise of Ignorance: Randomness as Lack of Certainty

"Quindi non avete una sola risposta alle vostre domande?"
"Adson, se l'avessi insegnerei teologia a Parigi."
"A Parigi hanno sempre la risposta vera?"
"Mai," disse Guglielmo, "ma sono molto sicuri dei loro errori."
(Umberto Eco: *Il Nome della Rosa*)

"So you don't have a unique answer to your questions?"
"Adson, if I had, I would teach theology in Paris."
"Do they always have a right answer in Paris?"
"Never", said William, "but there they are quite confident of their errors."
(Umberto Eco: *The Name of the Rose*)

After the lengthy technical introduction of the previous chapters, we are now ready to start estimating unknown quantities based on incomplete information and indirect observations. We adopt here the Bayesian point of view: Any quantity that is not known exactly, in the sense that a value can be attached to it with no uncertainty, is modeled as a random variable. In this sense, randomness means lack of certainty. The subjective part of this approach is clear: even if we believed that an underlying parameter corresponds to an existing physical quantity that could, in principle, be determined and therefore is conceptually deterministic, the lack of the *subject's* information about it justifies modeling it as a random variable. The question of whether a parameter will be modeled as a random variable is then answered according to how much we know about the quantity or how strong our beliefs are. This general guiding principle will be followed throughout the rest of the book, applied to various degrees of rigor.[1]

When employing statistical techniques to solve inverse problems, the notion of *parameter* needs to be extended and elaborated. In classical statistics, parameters are often regarded as tools, like, for example, the mean or the variance, which may be used to identify a probability density. It is not uncommon that even in that context,

---

[1] After all, as tempting as it may sound, we don't want to end up teaching theology in Paris.

parameters may have a physical[2] interpretation, yet they are treated as abstract objects. In inverse problems, parameters are more often *by definition* physical quantities, but they *are regarded* as statistical model parameters defining probability densities. Disquisitions about such differences in interpretation may seem unimportant, but these very issues often complicate the dialogue between statisticians and "inversionists."

Example 5.1 Consider the general inverse problem of estimating a quantity $x \in \mathbb{R}^n$ that cannot be observed directly, but for which indirect observations of a related quantity $y \in \mathbb{R}^m$ are available. We may, for example, be interested in the concentrations of certain chemical species (variable x) in a gas sample, but, for some reason, we cannot measure them directly; instead, we observe spectral absorption lines of light that passes through the specimen (variable y). A mathematical model describing light absorption by a mixture of different chemical compounds ties these quantities together. The fact that the variables that we are interested in are concentration values already carries *a priori* the information that they cannot take on negative values. In addition, by knowing where the sample is taken from, regardless of the subsequent measurement, may already give us a good idea of what to expect to be found in the gas sample. For instance, a sample taken from the Earth's atmosphere probably contains significant amounts of nitrogen and oxygen in approximately known concentrations. In fact, the whole process of measuring may be performed to confirm a hypothesis about the concentrations.

In order to set up the statistical framework, we need to express the distribution of y in terms of the parameter x. This is done by constructing the *likelihood model*. The design of the prior model will take care of incorporating any available *prior* information.

As the preliminary example above suggests, the statistical model for inverse problems comprises two separate parts:

1. The construction of the likelihood model;
2. The construction of the prior model,

both of which make extensive use of *conditioning* and *marginalization*. When several random variables enter the construction of a model, conditioning is one way to take into consideration one unknown at the time, assuming that the others are known. This allows us to construct complicated models step by step. For example, the joint density of x and y can be expressed in terms of the density of y assuming that x is known, multiplied by the density of x. This is the essence of the formula,

$$\pi_{XY}(x, y) = \pi_{Y|X}(y \mid x)\pi_X(x).$$

If, on the other hand, some of the variables appearing in the model are of no interest, we can eliminate them from the density by marginalizing them, that is by integrating them out. For example, if we have the joint density of three random variables

---

[2] The word "physical" in this context may be misleading in the sense that it makes us think of physics as a discipline. The use of the word here is more general, almost a synonym of "material" or "of observable nature," as opposed to something that is purely abstract.

$\pi_{XYV}(x, y, v)$, but we are not interested in V, we can marginalize V out by observing that

$$\pi_{XY}(x, y) = \int \pi_{XYV}(x, y, v)dv.$$

The parameter v of no interest is often referred to as noise, or as a nuisance parameter.

To set up the big picture, let X be a random variable representing the quantity of primary interest, and let Y denote a random variable representing the observable quantity. In the Bayesian approach to inverse problems, rather than trying to find a formula mapping Y to X, we ask what can be said about the possible values of X given that a realization y of Y is observed. The key to answer that question is Bayes' formula,

$$\pi_{X|Y}(x \mid y) = \frac{\pi_X(x)\pi_{Y|X}(y \mid x)}{\pi_Y(y)}, \quad y = \text{observed value of } Y.$$

To explain this formula in terms of the subjective probabilities discussed in Chap. 1, π_X encodes what we believe about X prior to observing Y, therefore we refer to it as prior density. The term $\pi_{Y|X}$ expresses the likelihood of a given observation $Y = y$, explaining why this term is referred to as likelihood. Finally, $\pi_{X|Y}$ expresses what we believe about X posterior to the observation, giving rise to the name posterior density. The formula above thus constitutes an updating scheme from "before" to "after." Ignoring the marginal value $\pi_Y(y)$ in the denominator, Bayes' formula is often summarized as

$$\text{posterior} \propto \text{likelihood} \times \text{prior.}$$

Adhering to the principle that statistics and probability are "common sense reduced to calculations," there is no universal prescription for the design of priors or likelihoods, although some recipes seem to be used more often than others. These will be discussed next.

5.1 Construction of Likelihood

From the point of view of inverse problems, the likelihood encodes the mathematical model connecting the unknown of primary interest to the observation, taking all possible uncertainties into consideration. It is in that frame of mind that we can think of the likelihood as answering the following question: *If we knew the unknown x and all other model parameters defining the data, how would the measurements be distributed?*

Since the construction of the likelihood starts from the assumption that, if $X = x$ were known, the measurement Y would be a random variable, it is important to understand the nature of its randomness. Randomness being synonymous of lack of certainty, we need to analyze what would make the data deviate from the predictions of our observation model. The most common sources of such deviations are

1. measurement noise in the data;
2. incompleteness of the observation model.

The probability density of the noise can, in turn, depend on unknown parameters, as will be demonstrated later through some examples. The second source of randomness is more complex, as it includes errors due to discretization, model reduction, and, more generally, all the shortcomings of a computational model, that is the discrepancy between the model and "reality"—in the heuristic sense of the word.[3]

5.2 Noise Models

We start by considering the construction of the likelihood based on different noise models, assuming that the forward model is completely known and a faithful description of reality.

5.2.1 Additive Noise

In inverse problems, it is very common to use additive models to account for measurement noise.

Assume that $x \in \mathbb{R}^n$ is the unknown of primary interest, that the observable quantity $y \in \mathbb{R}^m$ is ideally related to x through a functional dependence

$$y = f(x), \quad f : \mathbb{R}^n \to \mathbb{R}^m, \tag{5.1}$$

and that we are very certain of the validity of the model. The available measurement, however, is corrupted by noise, which we attribute to external sources or to instabilities in the measuring device, hence not dependent on x. Therefore, we write the *additive noise model*

$$Y = f(X) + E, \tag{5.2}$$

where $E : \Omega \to \mathbb{R}^m$ is the random variable modeling the noise. Observe that since X and Y are unknown, they are modeled as random variables—hence upper case letters—and (5.2) can be thought of as a *stochastic extension* of the deterministic model (5.1).

---

[3] Writing a model for the discrepancy between the model and reality implicitly, and arrogantly, assumes that we know the reality and we are able to tell how the model fails to describe it. Under these assumptions, the word "reality" is used in quotes, and should be understood as the "most comprehensive description available."

Let us denote the distribution of the error by

$$E \sim \pi_E(e).$$

Since we assume that the noise does not depend on X, fixing $X = x$ does not affect in any way the probability distribution of E. More precisely,

$$\pi_{E|X}(e \mid x) = \pi_E(e),$$

that is, the knowledge of X adds nothing to our knowledge of the noise. If, on the other hand, X is fixed, the only randomness in Y is due to E. Therefore,

$$\pi_{Y|X}(y \mid x) = \pi_E(y - f(x)), \tag{5.3}$$

that is, the randomness of the noise is translated by $f(x)$, as illustrated in Fig. 5.1.

In this setting, we assume that the distribution of the noise is known. Although this is a common assumption, in practice the distribution of the noise is seldom known exactly. More typically, the noise distribution itself may depend on parameters, collected in a variable θ. Writing the parametric model as

$$\pi_E(e) = \pi_E(e \mid \theta),$$

Equation (5.3) becomes

$$\pi_{Y|X}(y \mid x, \theta) = \pi_E(y - f(x) \mid \theta).$$

In anticipation of addressing the case in which the parameter θ may be poorly known, we include it explicitly in the formulas using the notation of conditional densities. At this point, however, θ is just a parameter in the likelihood model.

Fig. 5.1 Additive noise: the noise around the origin is shifted to a neighborhood of $f(x)$ without otherwise changing the distribution

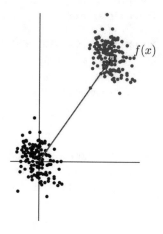

To illustrate the notation, assume that the noise E is zero-mean Gaussian with independent equally distributed components having variance σ^2, that is,

$$E \sim \mathcal{N}(0, \sigma^2 I_m),$$

where $I_m \in \mathbb{R}^{m \times m}$ is the identity matrix. The corresponding likelihood model is then

$$\pi_{Y|X}(y \mid x, \sigma^2) = \frac{1}{(2\pi)^{m/2} \sigma^m} \exp\left(-\frac{1}{2\sigma^2} \|y - f(x)\|^2\right),$$

with $\theta = \sigma^2$. If the noise variance is assumed known, we usually do not write the dependency explicitly, instead using the notation

$$\pi_{Y|X}(y \mid x) \propto \exp\left(-\frac{1}{2\sigma^2} \|y - f(x)\|^2\right),$$

and ignoring the normalizing constant altogether. If the variance is unknown, the dependency of the normalizing constant on it needs to be taken into account.

5.2.2   Multiplicative Noise

When discussing the additive noise, we formally solved for the noise E when $X = x$, and obtained a formula

$$E = Y - f(x) = \phi(Y),$$

that is, the two random variables E and Y are related through a simple one-to-one mapping comprising a shift by $f(x)$. The change of variable formula discussed in Sect. 1.2.4 yields the likelihood. Observe that here, the Jacobian of ϕ is identity: when the noise is multiplicative, this is no longer the case, as is illustrated in the following example.

Example 5.2   Consider a noisy amplifier that takes in a signal $f(t)$ and sends it out amplified by a constant factor $\alpha > 1$. We assume that the output signal is observed, and the input signal needs to be estimated.

If α is known and constant, the ideal model for the output signal is

$$g(t) = \alpha f(t), \quad 0 \le t \le T.$$

In practice, however, it may happen that the amplification factor is not constant but fluctuates slightly around a mean value α_0. To express that, we write a discrete likelihood model for the output by first discretizing the signal. Let

$$x_j = f(t_j), \quad y_j = g(t_j), \quad 0 = t_1 < t_2 < \cdots < t_n = T.$$

Assuming that the amplification at $t = t_j$ is a_j, we have a discrete model

$$y_j = a_j x_j, \quad 1 \le j \le n,$$

and, replacing the unknown quantities by random variables, we obtain the stochastic extension

$$Y_j = A_j X_j, \quad 1 \le j \le n,$$

which we write in vector notation as

$$Y = A \odot X, \tag{5.4}$$

the symbol $\odot$ denoting component-wise multiplication of the vectors $A, X \in \mathbb{R}^n$. Assume that A as a random variable is independent of X, as is the case, for instance, if the random fluctuations in the amplification are due to thermal phenomena. If A has the probability density

$$A \sim \pi_A(a),$$

to find the probability density of Y, conditioned on $X = x$, we fix $X = x$ and write

$$A_j = \frac{Y_j}{x_j}, \quad 1 \le j \le n.$$

Introduce the mapping $\phi : \mathbb{R}^n \to \mathbb{R}^n$,

$$\phi(y) = \begin{bmatrix} \frac{y_1}{x_1} \\ \vdots \\ \frac{y_n}{x_n} \end{bmatrix} = y \oslash x,$$

where the symbol $\oslash$ denotes componentwise division, whose Jacobian is

$$D\phi(y) = \begin{bmatrix} \frac{1}{x_1} & \\ & \ddots \\ & & \frac{1}{x_n} \end{bmatrix},$$

and conclude that

$$\pi_{Y|X}(y \mid x) = |\det(D\phi(y))|\pi_A(\phi(y)) = \frac{1}{x_1 x_2 \cdots x_n}\pi_A(y \oslash x). \tag{5.5}$$

Let us consider the case where all the variables are assumed to be positive, and A is *log-normally distributed*, i.e., the logarithm of A is normally distributed. For simplicity, we assume that the components of A are mutually independent, identically distributed. We write a stochastic model for A_j

$$A_j = e^{\mu + \sigma W_j}, \quad W_j \sim \mathcal{N}(0, 1),$$

and solving for W_j, we have

$$W_j = \frac{1}{\sigma}\left(\log A_j - \mu\right), \quad 1 \le j \le n,$$

or, in vectorial form,

$$W = \psi(A), \quad \psi_j(A) = \frac{1}{\sigma}\left(\log A_j - \mu\right).$$

For the change of variables formula, we need the Jacobian of ψ which is

$$D\psi(a) = \frac{1}{\sigma}\begin{bmatrix} \frac{1}{a_j} & & \\ & \ddots & \\ & & \frac{1}{a_n} \end{bmatrix}.$$

Noting that the distribution of W is

$$\pi_W(w) = \left(\frac{1}{2\pi}\right)^{n/2} \exp\left(-\frac{1}{2}\sum_{j=1}^{n} w_j^2\right),$$

and writing $\mu = \log a_0$, we obtain

$$\pi_A(a) = \left|\det\left(D\psi(a)\right)\right|\pi_W(\phi(a))$$

$$= \frac{1}{\sigma^n a_1 \cdots a_n}\left(\frac{1}{2\pi}\right)^{n/2} \exp\left(-\frac{1}{2\sigma^2}\sum_{j=1}^{n}(\log a_j - \mu)^2\right)$$

$$= \frac{1}{a_1 \cdots a_n}\left(\frac{1}{2\pi\sigma^2}\right)^{n/2} \exp\left(-\frac{1}{2\sigma^2}\sum_{j=1}^{n}\left(\log \frac{a_j}{a_0}\right)^2\right).$$

Consequently, the likelihood (5.5) with this particular noise model is

$$\pi_{Y|X}(y \mid x) = \left(\frac{1}{2\pi\sigma^2}\right)^{n/2}\frac{1}{x_1 \cdots x_n}\frac{1}{\frac{y_1}{x_1} \cdots \frac{y_n}{x_n}}\exp\left(-\frac{1}{2\sigma^2}\sum_{j=1}^{n}\left(\log \frac{y_j}{a_0 x_j}\right)^2\right)$$

$$= \left(\frac{1}{2\pi\sigma^2}\right)^{n/2}\frac{1}{y_1 \cdots y_n}\exp\left(-\frac{1}{2\sigma^2}\|\log\left(y\oslash(a_0 x)\right)\|^2\right),$$

where it is understood that the logarithm is applied component-wise.

5.2.3 Poisson Noise

In the previous examples, we started with an ideal deterministic model and corrupted the model adding independent noise. There are important applications where such error models are not appropriate, as in the following example where the forward model itself is intrinsically probabilistic.

Example 5.3 Assume that our measuring device is a photon collector, depicted schematically in Fig. 5.2 as a lens, and a counter of the photons emitted from N discrete sources, with average photon emission per observation time equal to x_j, $1 \leq j \leq N$. Our goal is to estimate the total emission from each source over a fixed time interval.

We model the photon collector by assuming that the total photon count is the weighted sum of the individual contributions. When the device is positioned above the jth source, it collects the photons from its immediate neighborhood. If we denote by a_k the weights of the contributions from sources located k positions away from the position of the device, the conditional expectation of the photons arriving to the counter is

$$\overline{y}_j = \mathrm{E}\big(Y_j \mid x\big) = \sum_{k=-L}^{L} a_k x_{j-k}.$$

The weights a_j are determined by the device and the index L is related to the width of the collector, as can be seen in Fig. 5.2. It is implicitly understood that $x_j = 0$ whenever $j < 1$ or $j > N$.

Considering the ensemble of all source points at once, we can express the conditional expectation in vector form

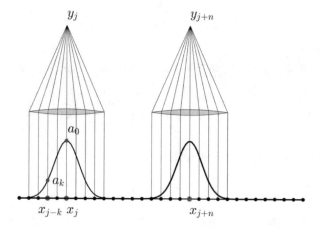

Fig. 5.2 The expected contributions from the different sources. The scalars a_k are the weights by which each source point $x_{j\pm k}$ contributes to the expected photon count when the center of the photon collector is above the jth point source

$$\bar{y} = E(Y \mid x) = Ax,$$

where $A \in \mathbb{R}^{N \times N}$ is the Toeplitz matrix

$$A = \begin{bmatrix} a_0 & a_{-1} & \cdots & a_{-L} & & & \\ a_1 & a_0 & & & \ddots & & \\ \vdots & & \ddots & & & a_{-L} & \\ a_L & & & \ddots & & & \vdots \\ & \ddots & & & & a_0 & a_{-1} \\ & & a_L & \cdots & & a_1 & a_0 \end{bmatrix}.$$

The parameter L defines the *bandwidth* of the matrix.

If the sources are weak emitters, the observation model just described is a photon-counting process. We may think that each Y_j is a Poisson process with mean $\bar{y}_j$

$$Y_j \mid x \sim \text{Poisson}((Ax)_j),$$

that is,

$$\pi_{Y_j \mid X}(y_j \mid x) = \sum_{k=0}^{\infty} \delta_k(y_j) \frac{(Ax)_j^k}{k!} \exp(-(Ax)_j).$$

Observe that, in general, there is no guarantee that the expectations $(Ax)_j$ are integers. If we assume that consecutive measurements are conditionally independent, the random variable $Y \in \mathbb{R}^N$ has probability density

$$\pi_{Y \mid X}(y \mid x) = \prod_{j=1}^{N} \pi_{Y_j \mid X}(y_j \mid x),$$

that is, the probability of a given outcome $y = (y_1, y_2, \ldots, y_N)$, given x, is

$$P\{Y = (y_1, y_2, \ldots, y_N) \mid x\} = \prod_{j=1}^{N} \frac{(Ax)_j^{y_j}}{y_j!} \exp(-(Ax)_j).$$

We express this relation simply by writing

$$Y \mid X \sim \text{Poisson}(Ax).$$

If we assume that the photon count is relatively large, the likelihood model can be approximated using the Gaussian approximation of the Poisson density discussed in Sect. 3.4.1, as

$$P\{Y = y \mid x\} \approx \prod_{j=1}^{N} \left(\frac{1}{2\pi(Ax)_j}\right)^{1/2} \exp\left(-\frac{1}{2(Ax)_j}(y_j - (Ax)_j)^2\right) \qquad (5.6)$$

$$= \left(\frac{1}{(2\pi)^N \det(C(x))}\right)^{1/2} \exp\left(-\frac{1}{2}(y - Ax)^{\mathsf{T}} C(x)^{-1}(y - Ax)\right),$$

where

$$C(x) = \text{diag}(Ax).$$

We point out that, for simplicity, we did not include the continuity correction in the formula above. The dependency on x is complicated here by the dependency of the covariance C on x.

It is instructive to see what the Poisson noise looks like, and to see how Poisson noise differs from Gaussian noise with constant variance. In Fig. 5.3, we have plotted a piecewise linear function $f : [0, 1] \rightarrow \mathbb{R}$, and assume that $f(t_j)$ at discrete points $t_j = j/N$ represent the mean values of mutually independent Poisson distributed random variables

$$X_j \sim \text{Poisson}(f(t_j)), \quad 0 \le j \le N.$$

For each j, we draw a random realization x_j and plot x_j versus t_j. From the plot, it is evident that the higher the mean, the higher the variance, in agreement with the fact that the mean and the variance are equal. By visual inspection, Poisson noise could be confused with *multiplicative noise* discussed in the previous section.

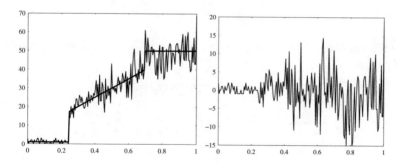

Fig. 5.3 Left panel: the average piecewise linear signal and a realization of a Poisson process, assuming that the values at each discretization point are mutually independent. Right panel: the difference between the noisy signal and the average

5.2.4 Composite Noise Models

Sometimes, the noise in the observation may comprise several different processes, each one contributing to the uncertainty in its own particular way. As an example elucidating this case, we assume that the photon counting convolution device discussed previously adds a noise component to the collected data. More precisely, consider an observation model of the form

$$Y = Z + E,$$

where $Z \sim \text{Poisson}(\mathsf{A}x)$ and $E \sim \mathcal{N}(0, \sigma^2 \mathsf{l})$.

To write the likelihood model, we begin by considering the conditional density of the jth component of Y, Y_j, assuming that $X = x$ and $Z_j = z_j$ are known, so that the only uncertainty arises from the additive noise. Then,

$$\pi_{Y_j | Z_j X}(y_j \mid z_j, x) \propto \exp\left(-\frac{1}{2\sigma^2}(y_j - z_j)^2\right),$$

where we have left out the normalizing constant that depends on the known variance σ^2. Observe that although x does not appear explicitly here, it is, in fact, a hidden parameter that affects the distribution of Z_j. It follows from the formula for the conditional probability density that

$$\pi_{Y_j Z_j | X}(y_j, z_j \mid x) = \pi_{Y_j | Z_j X}(y_j \mid z_j, x)\pi_{Z_j | X}(z_j \mid x),$$

where

$$\pi_{Z_j | X}(z_j \mid x) = \sum_{k=0}^{\infty} \frac{(\mathsf{A}x)_j^k}{k!}\exp\left(-(\mathsf{A}x)_j\right)\delta_k(z_j).$$

Since the value of z_j is not of interest here, we can marginalize it out. We obtain

$$\pi_{Y_j | X}(y_j \mid x) = \int \pi_{Y_j Z_j X}(y_j \mid z_j, x)\pi_{Z_j | X}(z_j \mid x)dz_j$$

$$= \sum_{k=0}^{\infty} \int \frac{(\mathsf{A}x)_j^k}{k!}\exp\left(-(\mathsf{A}x)_j\right)\pi_{Y_j | Z_j X}(y_j \mid z_j, x)\delta_k(z_j)dz_j$$

$$= \sum_{k=0}^{\infty} \frac{(\mathsf{A}x)_j^k}{k!}\exp\left(-(\mathsf{A}x)_j\right)\pi_{Y_j | Z_j X}(y_j \mid k, x)$$

$$\propto \sum_{k=0}^{\infty} \frac{(\mathsf{A}x)_j^k}{k!}\exp\left(-(\mathsf{A}x)_j\right)\exp\left(-\frac{1}{2\sigma^2}(y_j - k)^2\right),$$

which gives us the likelihood as a function of the variable of interest x. The joint likelihood $\pi_{Y|X}(y \mid x)$ is constructed in the same manner, and we see that

$$\pi_{Y|X}(y \mid x) = \prod_{j=1}^{n} \pi_{Y_j|X}(y_j \mid x) \qquad (5.7)$$

$$= \sum_{k \in \mathbb{N}^n} \frac{(Ax)_1^{k_1}}{k_1!} \cdots \frac{(Ax)_n^{k_n}}{k_n!} \exp\left(-\sum_{j=1}^{n}(Ax)_j\right) \exp\left(-\frac{1}{2\sigma^2}\|y - k\|^2\right),$$

where the sum extends over all n-dimensional lattice points of non-negative integers. Formula (5.7) states that the components Y_j are *conditionally independent*, that is, the components are mutually independent under the condition that X is known.

Notes and Comments

The likelihood plays the central role in the non-Bayesian statistical inference theory. Given the likelihood $\pi_{B|X}(b \mid x)$, the *maximum likelihood estimator* (MLE) is a value of x that maximizes the likelihood. In the non-Bayesian context, x is not a realization of a random variable, but rather a parameter, and MLE is the value that makes the observed value b most probable. Most importantly, the likelihood says nothing about the probability of x, and in fact, in this context, references to probabilities of x are meaningless. We don't pursue this argument in this book, but refer to any standard textbook on likelihood-based inference such as [65].

Chapter 6
Enter Subject: Construction of Priors

"The only relevant thing is uncertainty—the extent of our knowledge and ignorance. The actual fact of whether or not the events considered are in some sense determined, or known by other people, and so on, is of no consequence."

(Bruno deFinetti)

The prior density expresses what we know[1] about the unknown variable of interest before taking the measurements into account. The introduction of the prior is the reason why Bayesian inference is often characterized as being subjective. In reality, there are very few situations in which we do not have absolutely any idea about the unknown that we are trying to infer on. The level of a priori knowledge may vary a lot depending on how much previous experience with similar types of problems we have, and how deeply we are trusting our preconceived notions. For example, if the unknowns in question are concentrations, we know that they cannot take on negative values, and it would be foolish not to use that kind of knowledge to supplement the information coming from data. The saying that a specialist is someone who has made all possible mistakes in a field[2] is tantamount to saying that a specialist has acquired a lot of a priori knowledge by integrating past experiences on a particular topic, and uses the knowledge to avoid further mistakes. Willing or not, even in situations where there is no reason to have any expectation about the outcome of an event, it is very hard not to have a prior lurking in the background, which is why when the observed event is not in agreement with our latent expectations, we are surprised. In the lack of data, our inference is based solely on our prior. In the Bayesian framework, when

---

[1] Actually, it is not necessary that we *know*, it is sufficient that we *believe* in order to come up with a prior, since the prior is by its very nature subjective. The difference is subtle, as pointed out in Ludwig Wittgenstein's *On Certainty*, note 291: "We know that the earth is round. We have definitively ascertained that it is round. We shall stick to this opinion, unless our whole way of seeing nature changes. "How do you know that?"—I believe it."

[2] The quote is attributed to Niels Bohr in the form "An expert is a man who has made all the mistakes that can be made, in a narrow field."

© The Author(s), under exclusive license to Springer Nature Switzerland AG 2023
D. Calvetti and E. Somersalo, *Bayesian Scientific Computing*, Applied Mathematical Sciences 215, https://doi.org/10.1007/978-3-031-23824-6_6

the data become available and are taken into account, the prior is corrected in the light of them and the new description of the unknown is called the posterior. The proverbial wise man who changes his mind is definitely a Bayesian. Philosophically, declaring our a priori beliefs is a way to clarify what will be used to complement the information that can be extracted from the data. How much weight the prior has in guiding our estimation of the unknowns depends on the information content of the data: if the data are scarce or heavily contaminated by noise, the missing information will be provided by the prior which therefore may have significant influence on the estimate. If, on the other hand, the data are plentiful and highly informative, they will play a big role in the estimation of the unknown, thus decreasing the role of the prior. The next step is to address how to encode beliefs in the form of mathematical functions, a task that is particularly challenging when the knowledge is vague and of qualitative rather than quantitative type. Before addressing that question formally and systematically, we consider an example that illustrates how sometimes priors are used without being aware of them.

Example 6.1 Here is a banal example of hidden priors: assume that you want somebody to pick "a number, any number." If the somebody is a normal[3] human being, the answer may be 3, 7, 13—very popular choices in Judeo-Christian tradition—or 42—a favorite number in science fiction circles.[4] You may think that you did not have any prior on what to expect. Suppose now that the somebody you asked is a space alien in disguise who, taking your question very seriously, starts to recite: "40269167330954837...," continuing over a week. Although eventually the stream of digits will end, as the number is finite after all, you start to wonder if a lifetime is enough to finish a round of the game. Quite soon, probably much before the end of the week, you understand that this is not at all what you intended: the response is definitely in conflict with your prior, even though you were not aware that you had one. And in fact, the encounter with the alien made you realize that not only your prior was there, but it was quite strong too: there are infinitely many more numbers that takes more than a week to say—or your life time, or the lifetime of the universe, for that matter—than those you expected as an answer[5]!

Another, more pertinent example comes from the field of inverse problems.

Example 6.2 Assume that your new orthopedic surgeon, planning an operation of your knee that has caused problems, has asked you to bring at the next visit the MRI slides of your knee. On the phone, you ask the doctor what she expects to be the

---

[3] Preferably not from a math department, where people enjoy to come up with the most intricate answers for the simplest of questions, a tendency which becomes exacerbated when the answer involves numbers.

[4] In the subculture of mathematicians, even more exotic answers are likely, such as the Hardy–Ramanujan number, known also as the taxicab number, 1729. Essential here is to notice that to arrive at a number not more complex than this, it takes already the best number-minded brains.

[5] Jose Luis Borges has an elegant short story, *The book of sand*, about a book with an endless number of pages. Opening the book and reading the page number would be tantamount to choosing "any" number. Borges, possibly aware of the problem, and the intricacies related to the Axiom of Choice, leaves the page numbering a mystery.

problem, but being a new case, she says that she knows nothing yet about your case. On your way out of the house, you accidentally take the envelope with the MRI slides of the breast. As soon as the slides are taken out of the envelope, the doctor knows that you have grabbed the wrong slides, in spite of having just told you not to have any idea what to expect to see in the slides. Clearly, nobody is surprised of the comment, as "having no idea" or "knowing nothing" in this context is just a way to express the lack of information of some details.

The examples above clearly demonstrate how *extremely difficult it is to know nothing*.[6]

When an a priori belief is *qualitative* in nature and rather unreliable, the challenge becomes to find a *quantitative* formulation that is in agreement with its qualitative property and conveys the appropriate degree of uncertainty. For example, how can we express in a mathematical formula that we expect to see an MRI of the neck, or how we describe quantitatively what radiologists mean by "habitus of a malignant tumor"? Clearly, the problem is right at the edge between art and science. While the prior expresses our a priori belief about the unknown, subjective is by no means the same as arbitrary, as a prior position needs to be as defendable as any scientific statement. When a prior is relatively non-committal, it may be the case that the same data may support very different solutions, and excessive lack of commitment may be catastrophic.

6.1 Smoothness Priors

A popular class of priors for distributed parameters expresses a priori expectations about the smoothness properties of the solution.

We start by considering a one-dimensional signal $f(t)$ over a given time interval, whose values we do not know, and that we want to estimate from some indirect observations that are not specified here. Assume for simplicity that the time has been scaled so that the support of the signal is the interval $[0, 1]$. We divide the interval into n equal subintervals by lattice points t_j,

$$0 = t_0 < t_1 < \ldots < t_n = 1, \quad t_j = \frac{j}{n},$$

and seek to estimate the discretized values $x_j = f(t_j)$, $0 \le j \le n$. Consider the following descriptions of two different prior beliefs:

1. We know that $x_0 = 0$, and believe that the absolute value of the slope of f should be bounded by some $m_1 > 0$.
2. We know that $x_0 = x_n = 0$ and believe that the absolute value of the second derivative of f should be bounded by some $m_2 > 0$.

---

[6] Allegedly, Socrates said: "I am the wisest man alive, for I know one thing, and that is that I know nothing." Alas, the rest of us must admit that we know something.

Consider the first prior belief. By writing

$$f'(t_j) \approx \frac{x_j - x_{j-1}}{h}, \quad h = \frac{1}{n}, \quad 1 \leq j \leq n,$$

we may state the prior belief by saying that

$$x_0 = 0, \quad |x_j - x_{j-1}| \leq h\, m_1 \text{ with some uncertainty.} \tag{6.1}$$

To express the uncertainty, we introduce the random variables X_j, the values x_j representing their realizations. The boundary condition implies that $X_0 = 0$ with certainty, so we only need to write a probabilistic model for X_j, $1 \leq j \leq n$. We write a stochastic version of (6.1) as

$$X_j = X_{j-1} + \gamma W_j, \quad W_j \sim \mathcal{N}(0, 1), \quad \gamma = h\, m_1,$$

assuming that the random variables W_j are mutually independent. Collecting these conditions into a system of linear equations,

$$X_1 = X_1 - X_0 = \gamma W_1$$
$$X_2 - X_1 = \gamma W_2$$
$$\vdots \quad \vdots$$
$$X_n - X_{n-1} = \gamma W_n,$$

we arrive at the matrix formulation of the prior as

$$\mathsf{L}_1 X = \gamma W, \quad W \sim \mathcal{N}(0, \mathsf{I}_n),$$

where

$$\mathsf{L}_1 = \begin{bmatrix} 1 & & & \\ -1 & 1 & & \\ & \ddots & \ddots & \\ & & -1 & 1 \end{bmatrix} \in \mathbb{R}^{n \times n}, \quad X = \begin{bmatrix} X_1 \\ X_2 \\ \vdots \\ X_n \end{bmatrix}, \quad W = \begin{bmatrix} W_1 \\ W_2 \\ \vdots \\ W_n \end{bmatrix}.$$

The subindex of L_1 here refers to the prior being a first-order smoothness prior, related to the first-order derivative. This model is equivalent to specifying a Gaussian prior probability density,

$$X \sim \pi_X(x) \propto \exp\left(-\frac{1}{2\gamma^2} \|\mathsf{L}_1 x\|^2\right).$$

To find the covariance of the variable X, we first notice that since L_1 is an invertible matrix,

$$X = \gamma L_1^{-1} W,$$

hence

$$\text{cov}(X) = E(XX^{\mathsf{T}}) = \gamma^2 L_1^{-1} E(WW^{\mathsf{T}}) L_1^{-\mathsf{T}} = \gamma^2 L_1^{-1} L_1^{-\mathsf{T}}.$$

The inverse of the covariance matrix, or the *precision matrix*, can be expressed in terms of L_1 as

$$\text{cov}(X)^{-1} = \frac{1}{\gamma^2} L_1^{\mathsf{T}} L_1.$$

This prior is illustrated in the left panel of Fig. 6.1. Consider now the second prior model. By writing the finite difference approximation for the second derivative at the interior lattice points,

$$f''(t_j) \approx \frac{x_{j-1} - 2x_j + x_{j+1}}{h^2}, \quad 1 \le j \le n - 1,$$

the prior belief amounts to saying that

$$x_0 = x_n = 0, \quad |x_{j-1} - 2x_j + x_{j+1}| \le h^2 m_2,$$

with some uncertainty.

Introducing again the random variables X_j to model the signal at $t = t_j$, and assuming that the boundary conditions $X_0 = X_n = 0$ hold with certainty, a probabilistic model is needed only for the components X_j, $1 \le j \le n - 1$. We may express the prior belief as

$$X_j = \frac{1}{2}(X_{j-1} + X_{j+1}) + \frac{1}{2}\gamma W_j, \quad W_j \sim \mathcal{N}(0, 1), \quad \gamma = h^2 m_2,$$

or, equivalently, as a system of linear equations,

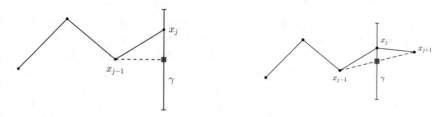

Fig. 6.1 Geometric interpretation of the first-order smoothness prior (left) and the second-order smoothness prior (right). The parameter γ expresses the standard deviation from the expected value marked by a red square, which is the previous value in the first-order prior model, and the average of the neighboring values in the second-order model

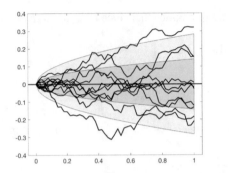 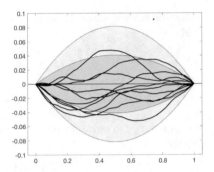

Fig. 6.2 Ten random draws from the priors corresponding to the first-order (left) and the second-order smoothness prior (right). The shaded areas indicate the credibility envelopes corresponding to one marginal standard deviation (darker) or two marginal standard deviations (lighter). In this example, we have set $n = 50$, $m_1 = 1$ and $m_2 = 2$

$$X_2 - 2X_1 = X_2 - 2X_1 + X_0 = \gamma W_1$$
$$X_3 - 2X_2 + X_1 = \gamma W_2$$
$$\vdots \quad \vdots$$
$$-2X_{n-1} - X_{n-2} = X_n - 2X_{n-1} + X_{n-2} = \gamma W_{n-1}.$$

The matrix form of this prior is

$$\mathsf{L}_2 X = \gamma W, \quad W \sim \mathcal{N}(0, \mathsf{I}_{n-1}),$$

where the subindex of L_2 reminds us that the prior is related to the second derivative, and

$$\mathsf{L}_2 = \begin{bmatrix} -2 & 1 & & \\ 1 & -2 & 1 & \\ & \ddots & \ddots & \\ & & 1 & -2 \end{bmatrix} \in \mathbb{R}^{(n-1)\times(n-1)}, \quad X = \begin{bmatrix} X_1 \\ X_2 \\ \vdots \\ X_{n-1} \end{bmatrix}, \quad W = \begin{bmatrix} W_1 \\ W_2 \\ \vdots \\ W_{n-1} \end{bmatrix}.$$

An illustration of this prior is shown in the right panel of Fig. 6.1. Having set up a prior model, a natural question is whether it reflects adequately the a priori assumptions that we had in mind. There are several ways to investigate the question. A natural one is to generate random draws from the prior and, if possible, to look at those realizations. The generation of random draws from a given generic distribution will be studied in greater detail later on, but for the priors just introduced, generating the samples is not particularly difficult. Since both prior models can be expressed in the form

$$\mathsf{L}_j X = \gamma W, \quad W \sim \mathcal{N}(0, \mathsf{I}), \quad j = 1, 2,$$

if we can generate a random sample from the standard white noise W,

$$\mathscr{W} = \left\{ w^{(1)}, w^{(2)}, \ldots, w^{(N)} \right\},$$

we can obtain an ensemble of realizations of X according to the formula

$$x^{(j)} = \gamma L_j^{-1} w^{(j)}, \quad 1 \leq j \leq N.$$

The sample $\mathscr{W}$ can be obtained using standard (pseudo)random number generators. Figure 6.2 shows 10 independent random draws from the probability densities corresponding to the two smoothness priors.

It is also useful to consider the uncertainties of each component x_j. One measure for the uncertainty is the *marginal variance*. Denoting by L either the matrix L_1 or L_2, and recalling that variances of single components of a multivariate random variables can be computed by using the joint probability density, we obtain

$$\begin{aligned}
\mathrm{var}(X_j) &= \int x_j^2 \pi_X(x)dx = \mathrm{cov}(X)_{jj} \\
&= \gamma^2 e_j^\mathsf{T} L^{-1} L^{-\mathsf{T}} e_j \\
&= \gamma^2 \| L^{-\mathsf{T}} e_j \|^2,
\end{aligned}$$

where e_j is the canonical jth coordinate vector. In Fig. 6.2, the uncertainties are visualized by shading the area between the values $\pm\sigma_j$ and $\pm 2\sigma_j$, where σ_j is the jth marginal standard deviation,

$$\sigma_j = \sqrt{\mathrm{var}(X_j)} = \gamma \| L^{-\mathsf{T}} e_j \|.$$

The envelopes are referred to as *credibility envelopes*.

6.1.1 Freeing the Boundary Values

The design of the smoothness prior above assumed that the boundary values of the unknown function were known and fixed at $x_0 = 0$ and, in the second-order case, also at $x_n = 0$. It is natural to ask whether the rather limiting boundary condition can be removed, in light of the fact that in many applications the boundary values are not known. The Bayesian philosophy provides a natural answer: If something is not known, it must be modeled as a random variable. For the sake of definiteness, consider the second-order smoothness prior, and write the discrete approximations for the second derivatives without assuming vanishing boundary conditions, but instead modeling the boundary values as independent Gaussian random variables. We have

$$X_0 = \alpha W_0$$
$$X_0 - 2X_1 + X_2 = \gamma W_1$$
$$X_1 - 2X_2 + X_3 = \gamma W_2$$
$$\vdots \quad \vdots$$
$$X_{n-2} - 2X_{n-1} + X_n = \gamma W_{n-1}$$
$$X_n = \alpha W_n$$

or, in matrix notation,

$$\mathsf{L}_\alpha X = \gamma W, \quad W \sim \mathcal{N}(0, \mathsf{I}_{n+1}),$$

where

$$\mathsf{L}_\alpha = \begin{bmatrix} \gamma/\alpha & & & & & \\ 1 & -2 & 1 & & & \\ & 1 & -2 & & & \\ & & & \ddots & & \\ & & & 1 & -2 & 1 \\ & & & & & \gamma/\alpha \end{bmatrix} \in \mathbb{R}^{(n+1)\times(n+1)}.$$

If we have some a priori belief about the range of the boundary values, we may use that information to set the parameter α. If no specific information is available, a possible way to set α is so that the marginal variances of the boundary values are close to those of the interior values. This can be accomplished by the following sequential adjustment strategy. First, assume that the boundary values are known to be zero, and compute the marginal variances as in the previous subsection. Let

$$\sigma_{\max} = \max\{\sigma_j\} = \max\{\gamma \|\mathsf{L}^{-\mathsf{T}} e_j\|\}.$$

We have seen that the maximum was attained in the middle of the interval: Assuming for simplicity that n is even, we have

$$\sigma_{\max} = \gamma \|\mathsf{L}^{-\mathsf{T}} e_{n/2}\|.$$

Setting $\alpha = \sigma_{\max}$ so that the marginal variances at the boundaries coincide with that in the center leads to

$$\mathsf{L}_\alpha = \begin{bmatrix} 1/\|\mathsf{L}_2^{-\mathsf{T}} e_{n/2}\| & & & & & \\ 1 & -2 & 1 & & & \\ & 1 & -2 & & & \\ & & & \ddots & & \\ & & & 1 & -2 & 1 \\ & & & & & 1/\|\mathsf{L}_2^{-\mathsf{T}} e_{n/2}\| \end{bmatrix}.$$

Fig. 6.3 Ten random draws from the prior corresponding to the second-order smoothness matrix L_α, together with the credibility envelopes corresponding to one and two marginal standard deviations

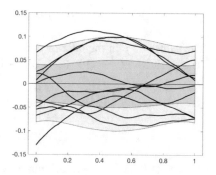

To test whether this procedure yields a prior reflecting our belief, we draw random samples from the prior thus obtained and plot the credibility envelopes. Figure 6.3 shows the results.

6.2 Generalization to Higher Dimensions

The smoothness priors just introduced in the one-dimensional case can be generalized to higher dimensions. Let f be a function defined over the rectangle $D = [0, 1] \times [0, 1]$ to be estimated through some indirect observation. Assume that the domain D is subdivided into n^2 equal squares, denoting the lattice points by

$$r_{jk} = (t_j, s_k), \quad t_j = \frac{j}{n}, \quad s_k = \frac{k}{n}, \quad 0 \le j, k \le n,$$

and the function values at the lattice points by $x_{jk} = f(r_{jk})$. A priori we believe that:

1. At the boundaries of D, f vanishes for sure, i.e.,

$$f\big|_{\partial D} = 0.$$

2. In the interior points, the absolute value of the Laplacian of f,

$$\Delta f = \frac{\partial^2 f}{\partial t^2} + \frac{\partial^2 f}{\partial s^2},$$

is bounded by some constant $m > 0$, with some uncertainty.

The next task is to express these prior beliefs in the form of a density for the multivariate random variable X.

Consider a matrix $x \in \mathbb{R}^{(n+1)\times(n+1)}$ containing the discrete values of f at the lattice points. We encode the belief that f vanishes at the boundary by setting

$$x_{0k} = x_{nk} = 0, \quad 0 \le k \le n, \quad x_{j0} = x_{jn} = 0, \quad 0 \le j \le n.$$

The values of f in the interior points will be modeled as random variables that we arrange into an $(n-1) \times (n-1)$ matrix,

$$\mathsf{X} = \begin{bmatrix} X_{11} & \cdots & X_{1,n-1} \\ \vdots & & \vdots \\ X_{n-1,1} & \cdots & X_{n-1,n-1} \end{bmatrix}.$$

We denote by $X^{(k)}$ the kth column of X, so that

$$\mathsf{X} = \begin{bmatrix} X^{(1)} & \cdots & X^{(n-1)} \end{bmatrix},$$

and the realizations of $X^{(k)}$ by $x^{(k)}$. Consider the second derivative of f with respect to the variable t, and its finite difference approximation

$$\frac{\partial^2 f}{\partial t^2}(x_{jk}) \approx \frac{x_{j-1,k} - 2x_{j,k} + x_{j+1,k}}{h^2}, \quad h = \frac{1}{n}. \tag{6.2}$$

To find a matrix notation, observe that the second derivative with respect to t at x_{jk} depends only on the values of f in the column k. By taking the vanishing boundary conditions into account, we may write the right-hand side of (6.2) using the matrix L_2 defined earlier,

$$\frac{x_{j-1,k} - 2x_{j,k} + x_{j+1,k}}{h^2} = \frac{1}{h^2}\left(\mathsf{L}_2 x^{(k)}\right)_j.$$

Consider now the second partial derivatives with respect to the variable s, and its finite difference approximation,

$$\frac{\partial^2 f}{\partial s^2}(r_{jk}) \approx \frac{x_{j,k-1} - 2x_{j,k} + x_{j,k+1}}{h^2}.$$

This time, we may express the right-hand side in terms of the $k-1$, k and $k+1$ columns of X,

$$\frac{x_{j,k-1} - 2x_{j,k} + x_{j,k+1}}{h^2} = \frac{1}{h^2}\left(x^{(k-1)} - 2x^{(k)} + x^{(k+1)}\right)_j.$$

To organize the computations efficiently, we define the operator stacking the columns of a matrix into a long column vector,

$$\text{vec} : \mathbb{R}^{(n-1)\times(n-1)} \to \mathbb{R}^{(n-1)^2}, \quad \text{vec}(X) = X = \begin{bmatrix} X^{(1)} \\ X^{(2)} \\ \vdots \\ X^{(n-1)} \end{bmatrix}. \tag{6.3}$$

Furthermore, we define the Kronecker product of two matrices as follows: If $A = [a_{ij}] \in \mathbb{R}^{p_1 \times q_1}$ and $B = [b_{ij}] \in \mathbb{R}^{p_2 \times q_2}$, their Kronecker product is the block matrix

$$A \otimes B = \begin{bmatrix} a_{11}B & \cdots & a_{1q_1}B \\ \vdots & & \vdots \\ a_{p_1 1}B & \cdots & a_{p_1 q_1}B \end{bmatrix} \in \mathbb{R}^{p_1 p_2 \times q_1 q_2},$$

where the (j, k)th block is a copy of the matrix B multiplied by the (j, k)th element of A. Kronecker products are very convenient for expressing finite difference approximations of the second partial derivatives in a compact fashion. Indeed, the vector containing the approximations of the second derivatives with respect to t can be expressed as

$$\begin{bmatrix} L_2 X^{(1)} \\ L_2 X^{(2)} \\ \vdots \\ L_2 X^{(n-1)} \end{bmatrix} = \begin{bmatrix} L_2 & & & \\ & L_2 & & \\ & & \ddots & \\ & & & L_2 \end{bmatrix} \begin{bmatrix} X^{(1)} \\ X^{(2)} \\ \vdots \\ X^{(n-1)} \end{bmatrix} = (I_{n-1} \otimes L_2)X.$$

Similarly, the vector of the approximations of the second derivatives with respect to s is written as

$$\begin{bmatrix} -2X^{(1)} + X^{(2)} \\ X^{(1)} - 2X^{(2)} + X^{(3)} \\ \vdots \\ X^{(n-3)} - 2X^{(n-2)} + X^{(n-1)} \\ X^{(n-2)} - 2X^{(n-1)} \end{bmatrix} = \begin{bmatrix} -2I_{n-1} & I_{n-1} & & & \\ I_{n-1} & -2I_{n-1} & I_{n-1} & & \\ & \ddots & \ddots & \ddots & \\ & & I_{n-1} & -2I_{n-1} & I_{n-1} \\ & & & I_{n-1} & -2I_{n-1} \end{bmatrix} \begin{bmatrix} X^{(1)} \\ X^{(2)} \\ \vdots \\ X^{(n-2)} \\ X^{(n-1)} \end{bmatrix}$$

$$= (L_2 \otimes I_{n-1})X.$$

The Kronecker matrix representations just introduced naturally yield the following finite difference approximation of the Laplacian:

$$\Delta f \approx \frac{1}{h^2} (I_{n-1} \otimes L_2 + L_2 \otimes I_{n-1}) X.$$

We are now ready to write the prior model conveying the belief that the Laplacian of f is bounded by a quantity m with some uncertainty,

$$(I_{n-1} \otimes L_2 + L_2 \otimes I_{n-1}) \, X = \gamma W,$$

where W is a standard normal random variable with independent components,

$$W \sim \mathcal{N}(0, I_{(n-1)^2}), \quad \gamma = h^2 m.$$

Random draws from the second-order smoothness prior can be generated by drawing the Gaussian random vectors w as realizations of W, and then solving the linear equation

$$(I_{n-1} \otimes L_2 + L_2 \otimes I_{n-1}) \, x = \gamma w, \quad w \sim \mathcal{N}(0, I_{(n-1)^2}).$$

Observe that while the finite difference matrix is rather large, it is also extremely sparse.

6.3   Whittle–Matérn Priors

The smoothness priors introduced in the previous section can be modified to account for a priori belief about the sizes of details in the unknown. We derive the corresponding flexible family of smoothness priors in the two-dimensional setting. Define a family of symmetric positive definite matrices of size $(n-1)^2 \times (n-1)^2$,

$$\mathsf{W}_\lambda = \frac{1}{\lambda^2} I_{(n-1)^2} - \Lambda,$$

where Λ is the discrete Laplacian approximation derived in the previous section,

$$\Lambda = I_{n-1} \otimes L_2 + L_2 \otimes I_{n-1}.$$

As λ increases, the matrix W_λ converges to the finite difference approximation of the Laplacian. On the other hand, if λ tends to zero, the scaled identity matrix becomes the dominant part, and in that case a random variable X satisfying the equation

$$\mathsf{W}_\lambda X = \gamma W, \quad W \sim \mathcal{N}(0, I_{(n-1)^2}), \tag{6.4}$$

is increasingly similar to a scaled version of the white noise W. This observation gives us a clue about the role of the parameter λ in controlling the granularity of the unknown as expressed by the prior corresponding to the above model. For this reason, λ is referred to as the *correlation length* parameter. We point out that the matrix W_λ is always positive definite, so it indeed defines a proper covariance matrix. We illustrate the effect of λ on the granularity in Fig. 6.4, where we show random draws computed according to Eq. (6.4) with different choices of λ.

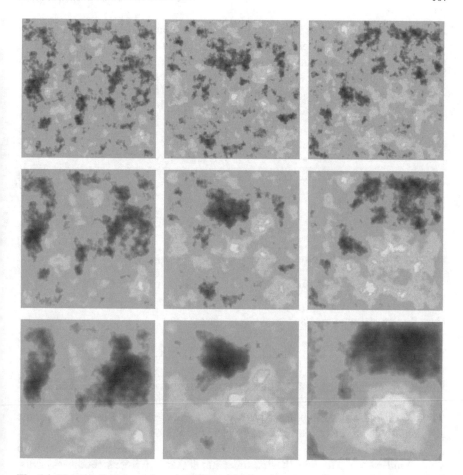

Fig. 6.4 Three random draws from the Whittle–Matérn prior with varying correlation lengths with $n = 150$. Denoting by $h = 1/n$, in the top row, $\lambda = 5\,h$; in the middle row: $\lambda = 10\,h$; and in the bottom row, $\lambda = 50\,h$. Observe that in each column, the right-hand side W in (6.4) was fixed. The dynamic ranges of the images can be controlled by the parameter γ

6.4 Smoothness Priors with Structure

The smoothness priors discussed in the previous sections can be enriched by adding structure to them. We start by considering the one-dimensional case.

Suppose that we believe a priori that the slope of the function f over the unit interval is bounded everywhere by a constant m except in a small interval around a known point $t^* \in (0, 1)$ where the absolute value of the slope can be larger by a multiplicative factor $\sqrt{\theta}, \theta > 1$. The reason for using the square root of the parameter as the multiplicative factor will become clear later, when we give an interpretation for this parameter. To build the prior, we discretize the interval by subdividing it into n subintervals and use finite difference approximation for the derivatives. Let $t_{j*} = j*/n$ be the point where the bound for the absolute value of the slope is scaled up. As in the previous section, we write the conditions defining the prior model in the form of a linear system

$$X_1 = X_1 - X_0 = \gamma W_1$$
$$X_2 - X_1 = \gamma W_2$$
$$\vdots \quad \vdots$$
$$X_{j*} - X_{j*-1} = \gamma\sqrt{\theta}W_{j*} \tag{6.5}$$
$$\vdots \quad \vdots$$
$$X_n - X_{n-1} = \gamma W_n,$$

and arrive at the matrix formulation

$$\mathsf{L}_1 X = \gamma \mathsf{D}_\theta^{1/2} W, \quad W \sim \mathcal{N}(0, \mathsf{I}_n). \tag{6.6}$$

Here D_θ is a diagonal matrix,

$$\mathsf{D}_\theta = \mathrm{diag}(1, \ldots, \theta, \ldots, 1) \in \mathbb{R}^{n \times n},$$

with θ in the $j*$th place on the diagonal, and the power $1/2$ in Eq. (6.6) means that a square root of the diagonal entries is taken. Similar construction can be made with the second-order smoothness prior, allowing the second derivative to become significantly larger in absolute value at selected locations.

The effect of this change can be seen in Fig. 6.5, showing ten random draws with $n = 50$ and $\sqrt{\theta} = 10$ for both the first-order and the second-order smoothness priors. The random draws from first order smoothness prior show that the realizations are allowed larger variability at the special point indicated by the dashed line; however, the prior does not force the realizations to jump: some of them pass the point without being affected by the modification. Similarly, in the case of the second-order smoothness, the kink corresponding to the higher curvature is not forced, and some of the realizations do not demonstrate any different behavior at the selected point.

After solving (6.6) for W,

$$W = \frac{1}{\gamma}\mathsf{D}_\theta^{-1/2}\mathsf{L}_1 X \sim \mathcal{N}(0, \mathsf{I}_{n-1}),$$

we conclude that the prior probability density is

$$\pi_X(x) \propto \exp\left(-\frac{1}{2\gamma^2}\|\mathsf{D}_\theta^{-1/2}\mathsf{L}_1 x\|^2\right),$$

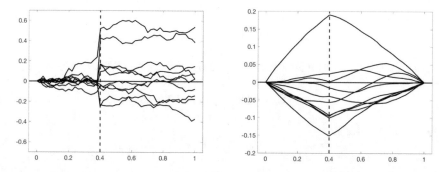

Fig. 6.5 Ten random draws from the modified first-order smoothness prior (left) and the second-order smoothness prior (right). The vertical line indicates the discretization point in which the absolute value of the first or the second derivative is allowed to take on a value ten times larger than elsewhere

ignoring, for the time being, the normalizing constant that depends on θ.

In our derivation, we assumed that the slope was allowed to be exceptionally different at one single location j^*. The procedure can be adapted to include as many exceptional points as we wish, leading to a richer class of prior densities.

6.5 Conditionally Gaussian Priors and Hierarchical Models

The next class of priors that we consider are particularly well suited to describe signals that are believed to be regular in a described sense almost everywhere, with a few occasional exceptions in behavior at locations that, unlike in the previous section, are unknown to us. In particular, not only we may not know where the anomalies are, but we may lack the information about how many of them there are.

For the sake of definiteness, let us consider the first-order smoothness model. We start, once again, with the smoothness prior model (6.5) and modify the standard deviations at all locations. Folding the factor γ into the variables θ_j, the new model becomes

$$X_1 = X_1 - X_0 = \sqrt{\theta_1}W_1$$
$$X_2 - X_1 = \sqrt{\theta_2}W_2$$
$$\vdots \quad \vdots$$
$$X_j - X_{j-1} = \sqrt{\theta_j}W_j \tag{6.7}$$
$$\vdots \quad \vdots$$
$$X_n - X_{n-1} = \sqrt{\theta_n}W_n,$$

or, in matrix form,

$$L_1 X = D_\theta^{1/2} W, \quad W \sim \mathcal{N}(0, I_n), \tag{6.8}$$

where D_θ is a diagonal matrix parametrized by a vector $\theta \in \mathbb{R}^n$ with non-negative entries,

$$D_\theta = \mathrm{diag}\,(\theta_1, \ldots, \theta_n)\,.$$

We solve (6.8) for W,

$$W = D_\theta^{-1/2} L_1 X,$$

and use the change of variables formula to find the probability density of X. In this case, however, we will not ignore the normalizing factor that depends on θ. The determinant of the Jacobian of the transformation above is

$$\det\left(D_\theta^{-1/2} L_1\right) = \det\left(D_\theta^{-1/2}\right)\det\left(L_1\right).$$

Both determinants on the right hand side of the equal sign are easy to evaluate: The matrix L_1 is lower triangular with ones on the diagonal, and $D_\theta^{-1/2}$ is diagonal, so

$$\det\left(D_\theta^{-1/2}\right) = \frac{1}{\sqrt{\theta_1 \theta_2 \cdots \theta_n}}, \quad \det\left(L_1\right) = 1.$$

Therefore, we conclude that the prior density of X, *given the parameter vector* θ, is

$$\pi_X\,(x \mid \theta) = \frac{1}{(2\pi)^{n/2}\sqrt{\theta_1 \theta_2 \cdots \theta_n}}\, \exp\left(-\frac{1}{2}\|D_\theta^{-1/2} L_1 x\|^2\right).$$

So far, we have assumed that the parameter vector θ is known and specified. But what happens if that is not the case, and θ is unknown? According to the Bayesian paradigm, if a parameter is not known to us, we model it as a random variable, and it must be estimated based on the data. Therefore, we introduce a random variable $\Theta \in \mathbb{R}^n$, and postulate that the prior just defined is, in fact, a conditional density, that is,

$$\pi_{X|\Theta}\,(x \mid \theta) = \frac{1}{(2\pi)^{n/2}\sqrt{\theta_1 \theta_2 \cdots \theta_n}}\, \exp\left(-\frac{1}{2}\|D_\theta^{-1/2} L_1 x\|^2\right)$$

$$\propto \exp\left(-\frac{1}{2}\|D_\theta^{-1/2} L_1 x\|^2 - \frac{1}{2}\sum_{j=1}^{n} \log \theta_j\right). \tag{6.9}$$

The price for introducing the new random variable Θ is that we need to specify our a priori belief about it in the form of a probability density π_Θ. More precisely, since now there are two random variables to estimate, we need to specify a prior for the pair (X, Θ). This is done in terms of the conditional prior for X given Θ and the prior for Θ, according to the formula

$$\pi_{X\Theta}(x, \theta) = \pi_{X|\Theta}(x \mid \theta)\pi_{\Theta}(\theta). \tag{6.10}$$

The model that we just derived is often referred to as *hierarchical prior model*, as it follows a layered hierarchy, proceeding from the unknown quantity of primary interest to a parameter defining its prior. The next task is to express in terms of this hierarchical prior model our belief about the primary unknown.

Let us get back to the original problem: We want to construct a prior for X conveying our belief that

1. The function f represented by the discrete values X_j varies little over most of the interval $[0, 1]$,
2. f may exhibit a few jumps at discrete points, although the number and locations are unknown.

Translating the above information into properties of Θ_j that control the jumps of X_j, we conclude that

a. The variables Θ_j should be positive,
b. Most of the variables Θ_j are presumably small,
c. A few of the components Θ_j can be large.

A way to make it possible and easy for the solution to have sudden jumps is to look for multivariate distributions with independent components, that is,

$$\pi_{\Theta}(\theta) = \prod_{j=1}^{n} \pi_{\Theta_j}(\theta_j).$$

Additionally, since there is no a priori preference for the location of the jumps, it is reasonable to assume that all components Θ_j are identically distributed. Moreover, the first observation requires that we look for a probability density that is supported on the positive real axis. In summary, we have reduced the search to probability distributions over the positive reals. The choice must now be guided by the remaining points.

In classical statistics, there are many one-dimensional probability distributions that favor small values while allowing outliers. Such distributions are often referred to as *fat-tailed densities* or *leptokurtic densities*. Our selection criterion takes into account computational convenience. Two distributions that stand out from that point of view are the *gamma distribution* and the *inverse gamma distribution*.

A random variable Θ_j follows the gamma distribution:

$$\Theta_j \sim \text{Gamma}(\beta, \theta_0), \quad \beta > 0, \ \theta_0 > 0,$$

if its probability density is given by

$$\pi_{\Theta_j}(\theta_j) = \frac{1}{\theta_0^{\beta}\Gamma(\beta)}\theta_j^{\beta-1}\exp\left(-\frac{\theta_j}{\theta_0}\right),$$

where $\Gamma(\beta)$ is the gamma function. The parameters β and θ_0, referred to as *hyperparameters*, are called the *shape* and *scale parameters*, respectively. Observe that when $\beta = 1$, we have the *Laplace distribution*,

$$\pi_{\Theta_j}(\theta_j) = \frac{1}{\theta_0} e^{-\theta_j/\theta_0}.$$

To gain a better understanding of the role of the hyperparameters, we consider the mean and the variance of the distribution. It can be shown that, for $\Theta_j \sim$ Gamma(β, θ_0),

$$\mathrm{E}\left(\Theta_j\right) = \beta\theta_0, \quad \mathrm{var}\left(\Theta_j\right) = \beta\theta_0^2.$$

Similarly, the hyperprior follows an inverse gamma distribution:

$$\Theta_j \sim \mathrm{InvGamma}\left(\beta, \theta_0\right), \quad \beta > 0, \ \theta_0 > 0,$$

if its probability density is

$$\pi_{\Theta_j}(\theta_j) = \frac{\theta_0^\beta}{\Gamma(\beta)} \theta_j^{-\beta-1} \exp\left(-\frac{\theta_0}{\theta_j}\right).$$

The change of variables formula reveals that if the random variable T is gamma-distributed, then $S = \theta_0^2/T$ follows the inverse gamma distribution. The mean and variance of the inverse gamma-distributed random variable are

$$\mathrm{E}\left(\Theta_j\right) = \frac{\theta_0}{\beta - 1}, \ \beta > 1, \quad \mathrm{var}\left(\Theta_j\right) = \frac{\theta_0^2}{(\beta - 1)(\beta - 2)}, \ \beta > 2.$$

Combining the conditionally Gaussian prior (6.9) and the gamma hyperprior yields the joint prior of the pair (X, Θ),

$$\pi_{X\Theta}(x, \theta) \propto \exp\left(-\frac{1}{2}\|\mathsf{D}_\theta^{-1/2}\mathsf{L}_1 x\|^2 - \frac{1}{2}\sum_{j=1}^n \log\theta_j\right) \prod_{j=1}^n \left(\theta_j^{\beta-1} \exp\left(-\frac{\theta_j}{\theta_0}\right)\right)$$

$$= \exp\left(-\frac{1}{2}\|\mathsf{D}_\theta^{-1/2}\mathsf{L}_1 x\|^2 - \frac{1}{2}\sum_{j=1}^n \log\theta_j + (\beta - 1)\sum_{j=1}^n \log\theta_j - \sum_{j=1}^n \frac{\theta_j}{\theta_0}\right)$$

$$= \exp\left(-\frac{1}{2}\|\mathsf{D}_\theta^{-1/2}\mathsf{L}_1 x\|^2 + \left(\beta - \frac{3}{2}\right)\sum_{j=1}^n \log\theta_j - \sum_{j=1}^n \frac{\theta_j}{\theta_0}\right).$$

A similar expression can be derived for the joint prior when the hyperprior is an inverse gamma distribution. We postpone any further discussion of hierarchical priors to later, when we will use them in connection with specific problems.

6.6 Sparsity-Promoting Priors

It is not uncommon to have the a priori belief that the unknown vector we are looking for has most of its components near zero, and that the data can be explained by only a few components substantially different from zero. Such solutions are called sparse, and we want to characterize priors that promote sparsity in the Bayesian framework. It goes without saying that setting most of the components equal to zero would reduce considerably the work burden, if it was not for the fact that usually we do not know a priori how many of the components are nonzero, and where the nonzero components are. As we will see later, the hierarchical priors introduced in the previous section are ideally suited to promote sparse solutions. Before addressing Bayesian sparsity in detail, we review some classic ways for promoting sparsity.

As a motivation, consider the following problem: Given a matrix $A \in \mathbb{R}^{m \times n}$ with $m < n$, and a vector $b \in \mathbb{R}^n$, consider the equation

$$b = Ax.$$

There are two possibilities: If $b \notin \mathcal{R}(A)$, that is, b is not in the range of A, no solutions can be found. On the other hand, if $b \in \mathcal{R}(A)$, there is at least one $x_0 \in \mathbb{R}^n$ such that the equation holds. However, since $m < n$, the null space of matrix A,

$$\mathcal{N}(A) = \{x \in \mathbb{R}^n \mid Ax = 0\},$$

has dimension at least $n - m$, therefore the equation above is satisfied also by all x such that

$$x \in H(b, A) = x_0 + \mathcal{N}(A) = \{x \in \mathbb{R}^n \mid x = x_0 + z, \ z \in \mathcal{N}(A)\}.$$

The set $H(b, A)$ is an affine subspace of $\mathbb{R}^n$ of dimension equal to that of the null space of A. After acknowledging the non-uniqueness of the solution, we may ask if there is a computationally efficient way to select a sparse solution among the many.

Before proceeding any further, we introduce some notations. The *support* of a vector $x \in \mathbb{R}^n$ is the set of indices corresponding to its nonzero components,

$$\text{supp}(X) = \{j \mid x_j \neq 0\} \subset \{1, 2, \ldots, n\}.$$

The *cardinality of the support*, that is, the number of nonzero entries in x, is called the "0-norm" of x,

$$\|x\|_0 = \text{card}\big(\text{supp}(x)\big) \in \{0, 1, \ldots, n\}.$$

A vector $x \in \mathbb{R}^n$ is *sparse*, if it has significantly fewer than n nonzero components,

$$\|x\|_0 \ll n.$$

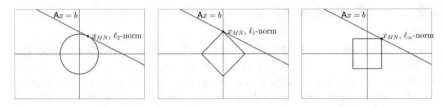

Fig. 6.6 A minimum-norm solution x_{MN} is the solution that lies on the affine space $Ax = b$ and is closest to the origin in terms of the given norm

How much fewer, however, is not a particularly well-defined measure. Usually, we recognize a sparse vector when we see one.[7]

After this preamble, we are ready to consider the standard minimum-norm selection principle, selecting the solution of smallest ℓ_2-norm,

$$x_{MN} = \operatorname{argmin}\{\|x\|_2 \mid x \in H(b, A)\}.$$

Geometrically, the minimum-norm solution corresponds to the point in $H(b, A)$ that lies closest to the origin in the Euclidean sense, see Fig. 6.6. It is natural to ask what happens if, instead of the Euclidean norm, other norms are used. The family of ℓ_p-norms constitute a natural generalization of the Euclidean norm,

$$\|x\|_p = \Big(\sum_{j=1}^{n} |x_j|^p\Big)^{1/p}, \quad 1 \le p < \infty, \tag{6.11}$$

augmented with the limiting case, the maximum norm, or ℓ_∞-norm,

$$\|x\|_\infty = \max\{|x_j|\} = \lim_{p \to \infty} \|x\|_p.$$

To understand what is the effect of changing the norm, recall what the ℓ_p-spheres look like, by plotting in $\mathbb{R}^2$ the unit spheres

$$\mathbb{S}_p = \{x \in \mathbb{R}^n \mid \|x\|_p = 1\}.$$

Figure 6.6 shows how the minimum-norm estimate changes if the ℓ_2-norm is replaced by ℓ_1 or ℓ_∞. The closest point to the origin in the ℓ_2-sense has both components x_1 and x_2 nonzero, while the point closest to the origin in the ℓ_1-sense is of the form $(0, x_2)$ with some $x_2 \ne 0$. In the two-dimensional case, this can be considered a sparse

---

[7] This is like the famous statement "I know it when I see it" of the Supreme Court Justice Potter Stewart discussing the threshold of obscenity.

solution. It is possible to imagine[8] that the same will hold in higher dimensions: the ℓ_1-sphere reaches out most strongly along the coordinate axes, and therefore the first contact with the affine subspace $H(b, A)$ would take place at a point where several components are zero.

The discussion above leads naturally to the following minimization problem:

$$\text{minimize } \|x\|_1 \text{ subject to } b = Ax.$$

Observe that for some data b, there may indeed be no x satisfying the linear constraint. However, we may always decompose b into two ℓ_2-orthogonal components,

$$b = b_1 + b_2, \quad b_1 \in \mathcal{R}(A), \quad b_2 \perp \mathcal{R}(A),$$

so that

$$\|b - Ax\|^2 = \|b_1 - Ax\|^2 + \|b_2\|^2 \geq \|b_2\|^2,$$

where the norm is the Euclidean ℓ_2-norm, and the best we can do is to require that x satisfies

$$Ax = b_1, \text{ implying that } \|b - Ax\|^2 = \|b_2\|^2,$$

prompting us to consider instead the modified minimization problem,

$$\text{minimize } \|x\|_1 \text{ subject to } \|b - Ax\|^2 = \|b_2\|^2.$$

This can be recast in the form of unconstrained minimization problem using Lagrange multipliers. In that formulation, we seek to find

$$x_\lambda = \text{argmin}\left\{\|x\|_1 + \lambda\left(\|b - Ax\|^2 - \|b_2\|^2\right)\right\},$$

where the Lagrange multiplier λ needs to be adjusted so that the constraint is satisfied. It is customary, however, to consider the equivalent minimization problem,

$$x_\alpha = \text{argmin}\left\{\|b - Ax\|^2 + \alpha\|x\|_1\right\}.$$

In the inverse problems literature, the solution is referred to as the ℓ_1-*penalized least squares solution*, while in the statistical literature it is called it the *least absolute shrinkage and selection operator*, or LASSO solution. The signal processing community refers to this problem as *basis pursuit denoising* (BPDN) problem.

From the point of view of Bayesian inference, we now have an excellent motivation to look for a prior that is concentrated on vectors with small ℓ_1-norm. The most straightforward selection, which is also one of the most popular ones,

---

[8] Well, it is possible, but not necessary easy, since our experience of the world is a three-dimensional affair. If we add time, and possibly with the help of colors, we can imagine four dimensions, but there is no way that we can stretch our imagination to envision five or higher dimensional objects.

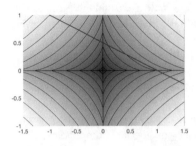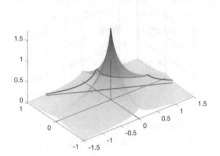

Fig. 6.7 Left: Equiprobability lines of the prior (6.12) with $p = 2/3$ and the affine subspace $H(b, \mathsf{A})$. Right: The profile of the posterior density along the affine space $H(b, \mathsf{A})$. Observe that the non-convexity of the ℓ_p-spheres implies that the prior density along the affine subspace $H(b, \mathsf{A})$ has more than one local maximum

$$\pi_X(x) \propto \exp\left(-\alpha \|x\|_1\right) = \exp\left(-\alpha \sum_{j=1}^n |x_j|\right),$$

is often referred to as ℓ_1-prior, and it heralded as the prototype of sparsity-promoting priors. Observe that the minimum ℓ_1-norm solution of the problem $\mathsf{A}x = b$ can be thought of as a problem of maximizing $\pi_X(x)$ along the affine space $H(b, \mathsf{A})$.

In the discussion above, we limited the parameter p to the interval $1 \le p \le \infty$. The discussion can be extended to the case $p < 1$, with the caveat that for $p < 1$, formula (6.11) does not define a proper norm, as the triangle inequality is not valid. Indeed, nothing prevents us from defining a prior of the form

$$\pi_X(x) \propto \exp\left(-\alpha \sum_{j=1}^n |x_j|^p\right), \quad 0 < p < 1. \tag{6.12}$$

This prior is even more sparsity-promoting than the ℓ_1-prior, which can be understood by considering the equiprobability contour lines corresponding to the ℓ_p-spheres, see Fig. 6.7: The contour lines reach out along the axes even more greedily than the corners of the ℓ_1-ball. However, the non-convexity of ℓ_p-balls for $p < 1$ implies that along the affine space $N(b, \mathsf{A})$, the prior π_X has typically more than one local maximum, posing a challenge for the optimization problem, see Fig. 6.7.

6.7   Kernel-Based Priors

Sometimes the unknown to be estimated represents a spatially distributed parameter, for example, the density of a material, and the prior should convey information of the size of the details we are expecting to see. To be more specific, assume that the unknown represents a physical property of subsurface structure of the Earth that we believe is horizontally layered, so that it is natural to represent it as a function of depth. Assume further that earlier geological investigations indicate that the stratified layers have a typical thickness, and we want to include this information into our prior. We may argue that values measured within a distance smaller than the typical layer thickness should be correlated, while the correlation over long distances should be minimal. Priors of this kind can be easily implemented using kernel methods.

Let $f : \mathbb{R}^d \to \mathbb{R}$ represent an unknown quantity to be estimated through an indirect observation, and define a *kernel function*,

$$K : \mathbb{R}^d \times \mathbb{R}^d \to \mathbb{R},$$

which is symmetric, that is,

$$K(r_1, r_2) = K(r_2, r_1) \text{ for all } r_1, r_2 \in \mathbb{R}^d.$$

Furthermore, we require that the kernel K satisfies the following property: For any set of points $r_1, r_2, \ldots, r_N \in \mathbb{R}^d$, with $r_j \neq r_k$ if $j \neq k$, the matrix

$$\mathsf{K} = \begin{bmatrix} K(r_1, r_1) & \cdots & K(r_1, r_N) \\ \vdots & & \vdots \\ K(r_N, r_1) & \cdots & K(r_N, r_N) \end{bmatrix} \in \mathbb{R}^{N \times N}$$

is symmetric positive semi-definite, that is, for any vector $v \in \mathbb{R}^N$, $v \neq 0$,

$$v^\mathsf{T} \mathsf{K} v = \sum_{j,k=1}^N v_j v_k K(r_j, r_k) \geq 0.$$

If K satisfies this property, it is said to be *a symmetric positive definite kernel*.[9] Before moving ahead, we introduce some explicitly symmetric positive definite kernels. Consider first a kernel that is separable,

$$K(r_1, r_2) = \varphi(r_1)\varphi(r_2),$$

for some function $\varphi : \mathbb{R}^d \to \mathbb{R}$. The matrix K in this case is a rank-one matrix,

$$\mathsf{K} = \begin{bmatrix} \varphi(r_1) \\ \varphi(r_2) \\ \cdots \\ \varphi(r_N) \end{bmatrix} \begin{bmatrix} \varphi(r_1) & \varphi(r_2) & \cdots & \varphi(r_N) \end{bmatrix} = \Phi \Phi^\mathsf{T},$$

---

[9] This nomenclature is confusing, as the matrices K are indeed required to be only positive semi-definite.

and, therefore, for any $v \neq 0$,

$$v^\mathsf{T}Kv = v^\mathsf{T}\Phi\Phi^\mathsf{T}v = \left(\Phi^\mathsf{T}v\right)^2 \geq 0.$$

The matrix K is only positive semi-definite, instead of positive definite, as the above quantity is zero for any vector v orthogonal to Φ. We may generalize the function K, and define

$$K(r_1, r_2) = \sum_{\ell=1}^{L} \varphi_\ell(r_1)\varphi_\ell(r_2).$$

If Φ_ℓ is a vector with entries $\varphi_\ell(r_j)$, then

$$v^\mathsf{T}Kv = \sum_{\ell=1}^{L} \left(\Phi_\ell^\mathsf{T}v\right)^2 \geq 0.$$

Moreover, if $L \geq N$ and the vectors Φ_ℓ are linearly independent, we may conclude that $v^\mathsf{T}Kv > 0$ for $v \neq 0$ or, equivalently, that the matrix is positive definite, while if $N > L$ the matrix is only positive semi-definite. Therefore, it is natural to consider kernels of the type

$$K(r_1, r_2) = \sum_{\ell=1}^{\infty} \varphi_\ell(r_1)\varphi_\ell(r_2), \tag{6.13}$$

assuming that the sum converges, and hope that for any N and for all choices of points r_j, the matrices K are symmetric positive definite. It turns out that, according to Mercer's theorem (see Notes and Comments at the end of the chapter), every positive definite kernel can be expressed as an infinite sum of the form (6.13). We shall not go any further in that direction, but instead present an example of kernel functions.

Example 6.3 Assume that we want to estimate a function f defined on a unit interval $[0, 1]$, which has been discretized by the nodes $0 = t_0 < t_1 < \ldots < t_n = 1$ with $t_j = j/n$, where we set $n = 100$. We denote by $x_j = f(t_j)$ the value of our unknown at t_j, and let the random variables X_j represent the grid values. We take the positive definite kernel function to be the Gaussian,

$$K(t, s) = \exp\left(-\frac{1}{2\gamma^2}(t - s)^2\right),$$

where the parameter $\gamma > 0$ controls the correlation length. One can show that the Gaussian kernel indeed is a positive definite kernel. It turns out that the covariance matrix C computed by using this kernel is not numerically positive definite, so we add to it a small multiple of the identity matrix,

$$C = \begin{bmatrix} K(t_0, t_0) & \cdots & K(t_0, t_n) \\ \vdots & & \vdots \\ K(t_n, t_0) & \cdots & K(t_n, t_n) \end{bmatrix} + \delta I_{n+1} = K + \delta I_{n+1},$$

where $\delta > 0$. The added scaled identity matrix guarantees that the matrix C is positive definite, since it follows from the positive definiteness of the kernel that

$$v^T C v = v^T K v + \delta \|v\|^2 \geq \delta \|v\|^2 > 0$$

for all $v \neq 0$. We define a prior model for X, setting

$$X \sim \mathcal{N}(0, C).$$

To test the effect of the correlation length parameter γ, we generate random draws from the prior as follows: Let

$$C = R^T R$$

be the Cholesky factorization of C. Then, we may write a whitened model for X as

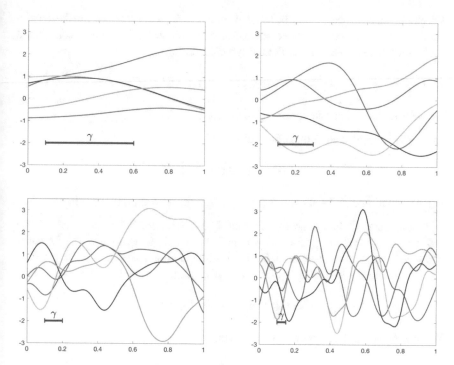

Fig. 6.8 Five random draws using the Gaussian kernel-based covariance matrix with varying correlation length parameter, $\gamma = 0.5$ (upper left), $\gamma = 0.2$, (upper right), $\gamma = 0.1$ (lower left) and $\gamma = 0.05$ (lower right). The number of discretization points in each case is $n = 101$

$$X = \mathsf{R}^{\mathsf{T}} W, \quad W \sim \mathcal{N}(0, \mathsf{I}_{n+1}),$$

providing a straightforward algorithm to generate random draws. Figure 6.8 shows five random draws from the prior using four different values for the correlation length parameters. The regularization parameter δ was chosen as $\delta = 10^{-8}$. The results show that γ controls well the sizes of the details in the signals.

Gaussian kernels are a popular choice also in higher dimensions,

$$K(r_1, r_2) = C \exp\left(-\frac{1}{2\gamma^2} \|r_1 - r_2\|^2\right),$$

where the parameter $\gamma > 0$ controls the width of the kernel and $C > 0$ is a scaling constant. Kernels depending on the distance of the points r_1 and r_2 are often referred to as *radial basis functions*. It can be shown that K is a positive definite kernel in $\mathbb{R}^n$.

6.8   Data-Driven Priors

Given a prior, its properties can be tested by drawing random samples from the density and, conversely, a representative sample of realizations can be used to construct a probability density. Given a sample of typical solutions,

$$\mathscr{S} = \{x^{(1)}, x^{(2)}, \ldots, x^{(p)}\}, \quad x^{(j)} \in \mathbb{R}^n,$$

we want to find a Gaussian prior π_X such that

(a) π_X is concentrated in the affine subspace

$$\mathscr{H} = \mathrm{span}\{x^{(1)}, \ldots, x^{(p)}\};$$

(b) π_X, restricted to the subspace $\mathscr{H}$, is distributed according to the sample.

If the subspace $\mathscr{H}$ is not the entire space $\mathbb{R}^n$ and the prior is supported on $\mathscr{H}$, we say that the prior is degenerate. This manifests itself as a covariance matrix that is symmetric positive semi-definite, but not positive definite.

To construct a non-degenerate Gaussian prior, we first arrange the data in a matrix,

$$\mathsf{X} = \begin{bmatrix} x^{(1)} & x^{(2)} & \cdots & x^{(p)} \end{bmatrix} \in \mathbb{R}^{n \times p}.$$

We then compute the mean value of the data,

$$\overline{x} = \frac{1}{p} \sum_{j=1}^{p} x^{(j)},$$

and define the centralized data matrix

$$\mathsf{X}_c = \left[x_c^{(1)} \; x_c^{(2)} \; \cdots \; x_c^{(p)} \right], \quad x_c^{(j)} = x^{(j)} - \overline{x}.$$

Using the centralized data matrix, we then compute the empirical covariance matrix,

$$\mathsf{C} = \frac{1}{p} \sum_{j=1}^{p} x_c^{(j)} \left(x_c^{(j)} \right)^\mathsf{T} = \frac{1}{p} \mathsf{X}_c \mathsf{X}_c^\mathsf{T}. \tag{6.14}$$

We observe that if C is not invertible, we cannot define the prior as a Gaussian as before. One way to construct an invertible covariance is to add a small perturbation,

$$\mathsf{C} \rightarrow \mathsf{C}_\delta = \mathsf{C} + \delta \mathsf{I}_n,$$

where $\delta > 0$ is a small regularization parameter, and postulate a prior model $X \sim \mathcal{N}(\overline{x}, \mathsf{C}_\delta)$. We demonstrate the idea by a computed example.

Example 6.4 Consider a face recognition problem in which we assume a priori that the unknown represents a photograph of a face. We construct a small database of publicly available black and white head shots.[10] After centering and cropping the faces, each image is a 145×201 grayscale matrix, the pixel values scaled so that they are in the range $[0, 1]$. From the database, we select the $p = 120$ representatives shown in Fig. 6.9. We rearrange the image matrices by using the vec-operator (6.3), organizing the data in a matrix $X \in \mathbb{R}^{29\,145 \times 120}$, compute the mean and the covariance matrix. Observe that the rank of the covariance matrix cannot exceed 120. Figure 6.10 shows the singular values of the matrix X, indicating that despite the strong redundancy of the data, almost all singular values are of significant size.

To test how well the prior thus defined represents the prior belief, we perform a number of random draws from the prior. To avoid computing the Cholesky decomposition of the huge covariance matrix C, instead we use the SVD of the centered data matrix,

$$\mathsf{X}_c = \mathsf{U}\mathsf{D}\mathsf{V}^\mathsf{T}.$$

Observing that there are $r \leq 120$ nonzero singular values, we may reduce the above representation to comprise only the r first singular vectors and singular values, and write

$$\mathsf{X}_c = \mathsf{U}_r \mathsf{D}_r \mathsf{V}_r,$$

---

[10] http://cvc.cs.yale.edu/cvc/projects/yalefaces/yalefaces.html. The database is known as Yale Faces, consisting of head shots of 15 individuals with different facial expressions, illumination, or with and without eyeware.

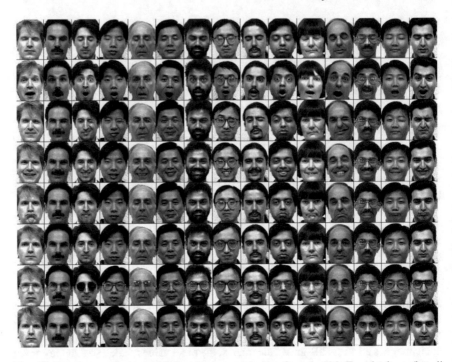

Fig. 6.9 Database consisting of 120 face images, originating from the Yale Face database, (http://cvc.cs.yale.edu/cvc/projects/yalefaces/yalefaces.html)

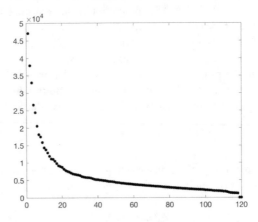

Fig. 6.10 Singular values of the face data matrix X. Observe that only the very last values are of negligible size

where $U_r \in \mathbb{R}^{n \times r}$ and $V_r \in \mathbb{R}^{p \times r}$ are obtained from U and V by retaining only the r first columns, and $D_r \in \mathbb{R}^{r \times r}$ is a diagonal matrix containing the non-vanishing singular values. We define the Gaussian random variable

$$Z = \bar{x} + \frac{1}{\sqrt{p}} U_r D_r W, \quad W \sim \mathcal{N}(0, I_r). \tag{6.15}$$

We observe that Z is a Gaussian random variable with mean $\bar{x}$ and covariance

$$\begin{aligned} E\big((Z - \bar{x})(Z - \bar{x})^\mathsf{T}\big) &= \frac{1}{p} U_r D_r E(WW^\mathsf{T}) D_r^\mathsf{T} U_r^\mathsf{T} \\ &= \frac{1}{p} U_r D_r D_r^\mathsf{T} U_r^\mathsf{T}, \end{aligned}$$

and by using the fact that columns of V_r are mutually orthogonal and $r \leq p$, we have

$$V_r^\mathsf{T} V_r = I_r,$$

implying that

$$\begin{aligned} \frac{1}{p} U_r D_r D_r^\mathsf{T} U_r^\mathsf{T} &= \frac{1}{p} U_r D_r V_r^\mathsf{T} V_r D_r^\mathsf{T} U_r^\mathsf{T} \\ &= \frac{1}{p} X_c X_c^\mathsf{T} = C. \end{aligned}$$

This calculation shows that we may use the random variable Z defined in (6.15) to generate random draws from the degenerate Gaussian distribution with the covariance matrix C. If we are interested in generating samples corresponding to the positive definite covariance matrix C_δ, we may define a random variable

$$Z_\delta = \bar{x} + \frac{1}{\sqrt{p}} U_r D_r W + \sqrt{\delta} W', \quad W \sim \mathcal{N}(0, I_r), \quad W' \sim \mathcal{N}(0, I_n),$$

where the variables W and W' are mutually independent. Finally, observe that by the definition of Z, we do not need to compute the full singular value decomposition of the matrix X_c which can be very time consuming and expensive. Instead, we only need to compute the first r singular vectors and singular values, which can be done economically, e.g., using in Matlab the routine svds.

Figure 6.11 shows the mean $\bar{x}$ of the prior, and 28 random draws of the variable Z, rearranged back into the form of 201×141 matrices and represented as grayscale images. We observe that the random draws bear similarities with some of the faces, and summarize rather well the original sample.

Notes and Comments

Smoothness priors have a long history, and they are rooted in classical regularization theory of inverse problems [20]. An extension of the idea of freeing the boundary values in the smoothness priors to dimensions higher than one can be found in

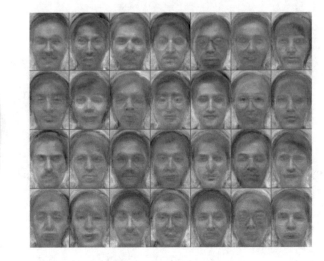

Fig. 6.11 The mean face (left) and 28 random draws from the Gaussian prior density, represented as images

[10, 16]. Whittle–Matèrn priors are popular in spatial statistics and geostatistics [54, 83], and they have recently been promoted in various applications to inverse problems [66, 67]. Interestingly, Whittle–Matérn priors can be expressed in an equivalent form in terms of kernel-based priors, where the kernel functions are certain modified Bessel functions. The advantage of defining them by using differential operators is that in this way, the matrices are sparse, which usually is not the case with kernel-based definitions. Finally, we point out that while we used finite difference approximations to discretize smoothness priors in rectangular domains, the methods are not restricted to simple domains. For more complicated domains and with different boundary values, finite element methods can be used.

Sparse inversion algorithms and sparsity-promoting priors have become particularly important in the wake of the boom in compressed sensing in image and signal processing [24, 27]. In this context, the ℓ_1-based optimization algorithm has become particularly popular [23]. A relatively recent review [37] contains a number of references to computational solutions to this optimization challenge. The ℓ_1-penalties for sparse reconstruction, however, has a longer history in signal processing and geophysics, as well as in statistics [76].

For a precise statement of Mercer's theorem, and its proof we refer to classical texts in functional analysis such as [62]. For more details about radial basis functions that play a significant role, e.g., in approximation theory, see, e.g., [7].

Chapter 7
Posterior Densities, Ill-Conditioning, and Classical Regularization

Now that the basic principles guiding the design of likelihoods and priors have been introduced, we are ready to welcome on the stage the main character in the Bayesian play of inverse problems, the posterior distribution, and in particular, the posterior density. Bayes' formula is the way in which prior and likelihood combine into the posterior density. In this chapter, we show through some examples how to explore and analyze posterior distributions. In later chapters, particular attention will be given to the design of numerical schemes of reduced complexity to deal with posteriors for high-dimensional inverse problems. In this chapter, we will build connections between posterior densities and classical regularization methods.

7.1 Likelihood Densities and Ill-Posedness of Inverse Problems

We start by considering the most popular types of posterior densities, corresponding to independent additive noise models. Let the observation model be of the form

$$b = f(x) + e,$$

where $f : \mathbb{R}^n \to \mathbb{R}^m$ is a known deterministic function, and $e \in \mathbb{R}^m$ represents additive observation noise. According to the Bayesian paradigm, all unknowns are modeled as random variables, leading to the stochastic extension,

$$B = f(X) + E.$$

© The Author(s), under exclusive license to Springer Nature Switzerland AG 2023
D. Calvetti and E. Somersalo, *Bayesian Scientific Computing*, Applied Mathematical Sciences 215, https://doi.org/10.1007/978-3-031-23824-6_7

If E and X are independent and have probability densities

$$E \sim \pi_E, \quad X \sim \pi_X,$$

according to Bayes' formula, the posterior distribution is

$$\pi_{X|B}(x \mid b) = \frac{\pi_{B|X}(b \mid x)\pi_X(x)}{\pi_B(b)}$$

$$\propto \pi_E\big(b - f(x)\big)\pi_X(x), \quad b = b_{\text{observed}}, \tag{7.1}$$

where the scaling factor in the denominator,

$$\pi_B(b) = \int \pi_{X,B}(x, b)dx = \int \pi_{B|X}(b \mid x)\pi_X(x)dx,$$

was assumed to be non-zero, and because it is simply a normalizing constant independent of x, it was ignored.

Inverse problems are usually *ill-posed*, meaning that a solution may not exist, or if it does, it may not be unique, and it tends to be highly sensitive to small errors in the data. All these disturbing features are reflected in the likelihood density, and can often be attenuated by using prior information, as the following simple example demonstrates.

Example 7.1 Consider the problem of determining a pair (x_1, x_2) of positive real numbers from noisy observations of their product, corresponding to an observation model

$$b = f(x) + e, \quad f(x) = x_1 x_2, \quad x_1, x_2 > 0.$$

The problem is clearly ill-posed in the sense described above, as the product of two numbers does not determine them in a unique way. To see how the ill-posedness affects the posterior, consider a plot of the likelihood density in the particular case in which e is a realization of a random variable E that is assumed to be scaled white noise,

$$E \sim \mathcal{N}(0, \sigma^2),$$

implying that

$$\pi_{B|X}(b \mid x) = \frac{1}{\sqrt{2\pi\sigma^2}}\exp\left(-\frac{1}{2\sigma^2}(b - x_1 x_2)^2\right).$$

Figure 7.1 shows that the likelihood density is a ridge-like function concentrated around the curve $x_1 x_2 = b$, and all values along that curve are equally likely solutions. The situation changes when we define an informative prior density, which in this case was chosen to be a Gaussian distribution. The prior density is informative in the sense that it contains complementary information about the unknown, and the posterior density, shown in the right panel, is well localized and leads to a rather good identification of a credible solution.

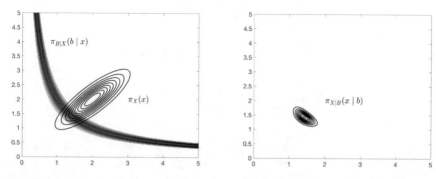

Fig. 7.1 Left: Likelihood density corresponding to a noisy observation of the product of two non-negative numbers, and an informative Gaussian prior. Right: Posterior density corresponding to this model

The previous example, in all its simplicity, highlights the importance of informative prior densities and is a clear warning about the danger of relaying exclusively on the likelihood. We elaborate further on this idea in connection with linear models.

Consider a linear observation model with additive noise,

$$b = Ax + e, \quad A \in \mathbb{R}^{m \times n},$$

and assume that the noise e is believed to be a realization of a scaled Gaussian white noise $E \sim \mathcal{N}(0, \sigma^2 I_m)$. This leads to a likelihood model of the form

$$\pi_{B|X}(b \mid x) \propto \exp\left(-\frac{1}{2\sigma^2}\|b - Ax\|^2\right).$$

Based on this model alone, what would be a good estimate for x? In the classical statistics, a frequently given answer is the *maximum likelihood (ML) estimator*,

$$x_{\mathrm{ML}} = \mathrm{argmax}\,\pi_{B|X}(b \mid x),$$

provided that such maximizer exists. For most inverse problems, the ML estimate is of little use due to the intrinsic ill-posedness of the problem. The ML estimate is a least squares solution,

$$x_{\mathrm{ML}} = \mathrm{argmin}\|b - Ax\|^2,$$

and can be expressed in terms of the components of the singular value decomposition

$$A = UDV^\mathsf{T}$$

of the matrix A. We write the orthogonal matrices U and V in terms of their orthonormal column vectors as

$$\mathsf{U} = \begin{bmatrix} u_1 & \cdots & u_m \end{bmatrix}, \quad \mathsf{V} = \begin{bmatrix} v_1 & \cdots & v_n \end{bmatrix}.$$

If A has r nonzero singular values $d_j > 0$, $1 \le j \le r$, we have

$$b - \mathsf{A}x = \sum_{j=1}^{m} u_j u_j^{\mathsf{T}} b - \sum_{j=1}^{r} u_j d_j v_j^{\mathsf{T}} x$$

$$= \sum_{j=1}^{r} u_j \left(u_j^{\mathsf{T}} b - d_j v_j^{\mathsf{T}} x \right) + \sum_{j=r+1}^{m} u_j u_j^{\mathsf{T}} b,$$

from which one can deduce that in order to minimize the norm of the residual, a least squares solution must be of the form

$$x_{\mathrm{ML}} = \sum_{j=1}^{r} \frac{u_j^{\mathsf{T}} b}{d_j} v_j + \sum_{j=r+1}^{n} \alpha_j v_j, \tag{7.2}$$

where the coefficients α_j can be chosen arbitrarily. The second term of the right-hand side of (7.2) is a general vector in the null space of the matrix A, and has no effect on the observation and, conversely, the observation does not contain any information about it. We therefore observe that, if $r < n$, hence the rank of the matrix of A is less than n, the ML estimate cannot be unique. Furthermore, it is usually very sensitive to noise, because any noise component along the singular vector u_j is divided by d_j, and the smaller d_j, the larger the amplification of the noise.

More generally, if the noise is Gaussian with SPD covariance matrix Σ, the likelihood density is of the form

$$\pi_{B|X}(b \mid x) \propto \exp\left(-\frac{1}{2}(b - \mathsf{A}x)^{\mathsf{T}} \Sigma^{-1}(b - \mathsf{A}x) \right),$$

and the maximum likelihood estimator is the minimizer of the quadratic form

$$\mathcal{E}_0(x, b) = \frac{1}{2}(b - \mathsf{A}x)^{\mathsf{T}} \Sigma^{-1}(b - \mathsf{A}x).$$

By the positive definiteness of Σ, it admits a symmetric factorization , for example, the Cholesky decomposition, or, the square root,

$$\Sigma = \Sigma^{1/2} \Sigma^{1/2},$$

allowing us to write

$$\mathcal{E}_0(x, b) = \frac{1}{2}\|b' - \mathsf{A}'x\|^2, \quad b' = \Sigma^{-1/2}b, \quad \mathsf{A}' = \Sigma^{-1/2}\mathsf{A}.$$

The normal equations that a least squares solution needs to satisfy can be written as

$$(A')^{\mathsf{T}}(A')x = (A')^{\mathsf{T}}b'$$

or,

$$A^{\mathsf{T}}\Sigma^{-1}Ax = A^{\mathsf{T}}\Sigma^{-1}b.$$

This equation has a unique solution if and only if the matrix $A^{\mathsf{T}}\Sigma^{-1}A$ is invertible, which usually is not guaranteed.

7.2 Maximum a Posteriori Estimate and Regularization

The classical deterministic approach for the solution of ill-posed problems is to approximate them with a nearby well-posed problem, an approach generally called *regularization*. Given an observation model with additive noise,

$$b = f(x) + e,$$

the idea behind regularization is to replace the functional,

$$\mathcal{E}_0(x, b) = \|b - f(x)\|^2,$$

whose minimizer is ill-defined, with the functional

$$\mathcal{E}_\alpha(x, b) = \|b - f(x)\|^2 + \alpha\mathcal{R}(x), \tag{7.3}$$

where $\mathcal{R}$ is a suitably chosen *regularization functional*, and $\alpha > 0$ is a *regularization parameter* that determines the relative weights of the two terms. We denote the minimizer of (7.3), if it exists, by x_α.

The selection of the value of the regularization parameter for α is crucial for the quality of the solution. One of the most popular selection criteria is based on *Morozov's discrepancy principle* that can be summarized as follows. If an estimate for the norm of the additive noise is available,

$$\|e\| \approx \eta,$$

the value of the parameter α is implicitly defined by the condition

$$\|b - f(x_\alpha)\| = \eta.$$

The motivation behind Morozov discrepancy principle is that any vector satisfying the above condition is acceptable and compatible with the noise level, while a residual norm smaller than the norm of the noise would indicate that the solution is fitting to the noise, which is usually a bad idea, in particular, for ill-posed problems. The process of finding the minimizer and adjusting the value of the regularization parameter is called *Tikhonov regularization*.

The selection of the regularization functional is not a trivial problem, because $\mathcal{R}$ implicitly determines the undesirable features of the candidate solutions that they should be penalized for. The general principle for the design of the regularization functional is that it will favor solutions for which $\mathcal{R}(x)$ is small. Popular regularization functionals are of the form

$$\mathcal{R}(x) = \|\mathsf{R}x\|^2,$$

where $\mathsf{R} \in \mathbb{R}^{k \times n}$ is a properly chosen matrix. Often, the matrix R is chosen to be invertible, although this is not a necessary condition. Observe that for any vector in the null space of R, there is no penalty, indicating that the regularization favors solutions that are in or near the null space of R.

To establish a connection between classical regularization methods and the Bayesian approach, assume that the regularization functional is chosen so that the expression

$$\mu_X(x) \propto \exp\left(-\frac{1}{2\theta}\mathcal{R}(x)\right), \quad \theta > 0, \tag{7.4}$$

defines a probability distribution, playing a role analogous to that of the prior. Under the assumption that the additive error is scaled white noise, that is, e is a realization of $E \sim \mathcal{N}(0, \sigma^2 \mathsf{I}_m)$, the posterior distribution becomes

$$\pi_{X|B}(x \mid b) \propto \exp\left(-\frac{1}{2\sigma^2}\|b - f(x)\|^2 - \frac{1}{2\theta}\mathcal{R}(x)\right).$$

Similarly to the process leading to the maximum likelihood estimator, we define the *maximum a posteriori (MAP) estimate* as

$$x_{\mathrm{MAP}} = \mathrm{argmax}\, \pi_{X|B}(x \mid b).$$

In the particular case that we are considering, computing the MAP estimate amounts to finding a minimizer of the negative logarithm of the posterior,

$$x_{\mathrm{MAP}} = \mathrm{argmin}\left\{\|b - f(x)\|^2 + \frac{\sigma^2}{\theta}\mathcal{R}(x)\right\}.$$

Therefore, in this example, the Tikhonov regularized solution x_α coincides with the x_{MAP} estimate, if we set the regularization parameter $\alpha = \sigma^2/\theta$.

A natural question is whether there is a Bayesian analogue of Morozov's discrepancy principle. We observe that setting the prior according to (7.4) assumes that θ is known, and adjusting θ is tantamount to adjusting our prior belief, an operation that should be independent of the observations, hence of the noise level. However, it is quite common to not have a priori a set value for θ in mind. In that case, according to the Bayesian paradigm, the parameter θ should be modeled as a random variable, in which case we are in the case of the hierarchical priors discussed in the previous chapter. Thus, we write the prior as a conditional prior,

$$\mu_{X|\Theta}(x \mid \theta) = N(\theta)\exp\left(-\frac{1}{2\theta}\mathcal{R}(x)\right), \tag{7.5}$$

where $N(\theta)$ is a normalizing factor that depends on the parameter, and then introduce a hyperprior for Θ. Observe that the normalizing constant may, in general, be difficult to evaluate, as it depends on the functional form of $\mathcal{R}(x)$.

We close this chapter with an example that motivates the need for sampling methods and underlines the importance of finding effective samplers.

Example 7.2 Consider the simple inverse problem of estimating $x \in \mathbb{R}^2$ from the noisy observation

$$b = x_1^3 - 2x_2 + e,$$

where e is a realization of a scaled white noise, $E \sim N(0, \sigma^2)$ with $\sigma = 0.1$. For the sake of definiteness, assume that

$$b = b_{\text{observed}} = 0.1.$$

Furthermore, assume that the prior is, up to a normalizing constant,

$$\pi_X(x) \propto e^{-\|x\|},$$

By Bayes' formula, we have that

$$\pi_{X|B}(x \mid b) \propto p(x) = \exp\left(-\|x\|\right)\exp\left(-\frac{1}{2\sigma^2}(b - (x_1^3 - 2x_2))^2\right).$$

To have an idea of what the posterior density looks like, we set up a rejection sampling scheme. The plan is to

1. Draw randomly from the prior density;
2. Accept or reject the proposal based on the likelihood.

Therefore, we need to find first a way to sample from the prior. To this end, we transform the prior density in polar coordinates,

$$x_1 = \varphi_1(r, \theta) = r\cos\theta, \quad x_2 = \varphi_2(r, \theta) = r\sin\theta.$$

The determinant of the Jacobian of this transformation is

$$\det\left(D\varphi(r, \theta)\right) = r,$$

and therefore, we have

$$\pi_{R\Theta}(r, \theta) = Ce^{-r}r,$$

where C is a normalizing constant. To find the value of this constant, we integrate the expression over the whole space, getting

$$\int_0^{2\pi} \int_0^{\infty} \pi_{R\Theta}(r, \theta) dr d\theta = C \int_0^{2\pi} \int_0^{\infty} e^{-r} r \, dr d\theta = 2\pi C,$$

implying that

$$C = \frac{1}{2\pi}.$$

In the rejection sampling algorithm, we set

$$q(x) = \frac{1}{2\pi} e^{-\|x\|}.$$

We then have

$$p(x) = \exp\left(-\|x\|\right) \exp\left(-\frac{1}{2\sigma^2}(b - (x_1^3 - x_2))^2\right) \leq e^{-\|x\|} = 2\pi q(x),$$

so the condition guaranteeing that the rejection sampling works is valid for $M = 2\pi$. The acceptance ratio in this case becomes

$$\alpha = \frac{p(x)}{2\pi q(x)} = \exp\left(-\frac{1}{2\sigma^2}(b - (x_1^3 - 2x_2))^2\right) \leq 1.$$

We therefore need only to find a way to draw from the prior. However, this is relatively straightforward in polar coordinates,

$$\pi_{R,\Theta}(r, \theta) = \frac{1}{2\pi} e^{-r} r,$$

allowing us to draw R and Θ separately. The angular variable is drawn from uniform distribution

$$\Theta \sim \text{Uniform}([0, 2\pi]).$$

To draw R, we write the cumulative distribution,

$$\Phi_R(r) = \int_0^r e^{-t} t \, dt = 1 - (1 + r)e^{-r},$$

and use the algorithm discussed in Sect. 4.5. Let t be a realization of $T \sim \text{Uniform}([0, 1])$. To obtain a random draw of R, we need to solve the equa-

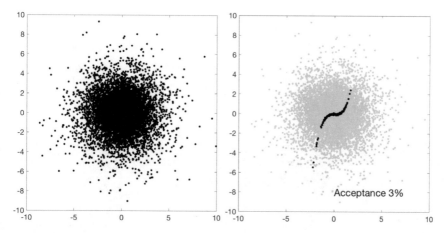

Fig. 7.2 Left: A sample of 10 000 points drawn from the prior density. Right: Rejection sampling result, the accepted points plotted as black dots. In this example, only about 3% of the prior sample was accepted

tion

$$\Phi_R(r) = t, \text{ or } (1 + r)e^{-r} = 1 - t.$$

This can be done numerically, e.g., by using the Newton algorithm for non-linear equations.

The result of the rejection sampling is shown in Fig. 7.2. Observe that the likelihood density is a narrow ridge around the curve $x_2 = 2x_1^3 - b$, and only few of the prior points fall on this ridge, leading to acceptance. This example highlights the need for more efficient sampling methods than the rejection sampling.

Notes and Comments

The connections between Tikhonov regularization and Bayesian inverse problems have been elaborated further in references [20, 49]. As pointed out here, the matrix R defining the penalty functional needs for not be invertible. A sufficient condition for making the regularized problem well-posed in the case of a linear forward model is that $\mathcal{N}(A) \cap \mathcal{N}(R) = \{0\}$, thus leaving no ambiguity in the solution.

It is not uncommon that a regularization term $\mathcal{R}(x) = \|x\|^2$ is called Tikhonov regularization, and $\mathcal{R}(x) = \|Rx\|^2$ generalized Tikhonov regularization. This nomenclature does not do justice to Tikhonov, who considered even more general penalty functions. We refer here to [77–79] for further details.

Chapter 8
Conditional Gaussian Densities

Gaussian probability distributions are the workhorse in Bayesian scientific computing, providing a well understood subclass of distributions that often allow closed form solutions to inference problems. Not only do the Gaussian distributions play a role similar to that of linear operators in analysis and in inverse problems in particular, as pointed out earlier, but there is an even deeper connection between these two classes, as seen in this chapter.

8.1 Gaussian Conditional Densities

We start by deriving the general form of conditional distributions for jointly Gaussian random variables.

Let $Z \in \mathbb{R}^N$ be a Gaussian random variable, and assume that its first n components are unknown, $0 < n < N$, while the remaining $m = N - n$ ones are observed. From the notion of independence, it is clear that if the two groups of components are independent, observing the latter gives no information about the former, while in the general case inference through correlation will occur, as shown below.

To write the density of the n first components of Z conditioned on knowledge of the last m ones, we start with a partitioning of Z of the form

$$Z = \begin{bmatrix} Z_1 \\ Z_2 \end{bmatrix} \begin{matrix} \in \mathbb{R}^n \\ \in \mathbb{R}^m \end{matrix}, \quad n + m = N,$$

and we express the probability density of Z as the joint density of Z_1 and Z_2,

$$\pi_Z(z) = \pi_{Z_1, Z_2}(z_1, z_2).$$

© The Author(s), under exclusive license to Springer Nature Switzerland AG 2023
D. Calvetti and E. Somersalo, *Bayesian Scientific Computing*, Applied Mathematical
Sciences 215, https://doi.org/10.1007/978-3-031-23824-6_8

The probability density of Z_1, under the condition that Z_2 is known, is given by

$$\pi_{Z_1|Z_2}(z_1 \mid z_2) \propto \pi_{Z_1,Z_2}(z_1, z_2), \quad z_2 = z_{2,\text{observed}}.$$

Observe that if we are investigating a Gaussian linear model with additive noise, and the model is

$$Z_1 = AZ_2 + E,$$

the conditional density above corresponds to the likelihood, while reversing the roles of Z_1 and Z_2 yields the posterior density.

In the following, we assume that $Z \in \mathbb{R}^N$ is a zero mean Gaussian random variable with symmetric positive definite covariance matrix $\mathbf{C} \in \mathbb{R}^{N \times N}$, the joint probability density of $Z_1 \in \mathbb{R}^n$ and $Z_2 \in \mathbb{R}^m$ being

$$\pi_{Z_1,Z_2}(z_1, z_2) \propto \exp\left(-\frac{1}{2} z^{\mathsf{T}} \mathbf{C}^{-1} z\right). \tag{8.1}$$

To investigate how this expression depends on z_1 when z_2 is given, we start by partitioning the covariance matrix,

$$\mathbf{C} = \begin{bmatrix} \mathbf{C}_{11} & \mathbf{C}_{12} \\ \mathbf{C}_{21} & \mathbf{C}_{22} \end{bmatrix} \in \mathbb{R}^{N \times N}, \tag{8.2}$$

where

$$\mathbf{C}_{11} \in \mathbb{R}^{n \times n}, \quad \mathbf{C}_{22} \in \mathbb{R}^{m \times m},$$

and by the symmetry of $\mathbf{C}$,

$$\mathbf{C}_{12} = \mathbf{C}_{21}^{\mathsf{T}} \in \mathbb{R}^{n \times m}.$$

We denote the precision matrix $\mathbf{C}^{-1}$ by $\mathbf{B}$, and we partition it according to the partition of $\mathbf{C}$,

$$\mathbf{C}^{-1} = \mathbf{B} = \begin{bmatrix} \mathbf{B}_{11} & \mathbf{B}_{12} \\ \mathbf{B}_{21} & \mathbf{B}_{22} \end{bmatrix} \in \mathbb{R}^{N \times N}. \tag{8.3}$$

We then write the quadratic form appearing in the exponent of (8.1) as

$$z^{\mathsf{T}} \mathbf{B} z = z_1^{\mathsf{T}} \mathbf{B}_{11} z_1 + 2 z_1^{\mathsf{T}} \mathbf{B}_{12} z_2 + z_2^{\mathsf{T}} \mathbf{B}_{22} z_2 \tag{8.4}$$

$$= \left(z_1 + \mathbf{B}_{11}^{-1} \mathbf{B}_{12} z_2\right)^{\mathsf{T}} \mathbf{B}_{11} \left(z_1 + \mathbf{B}_{11}^{-1} \mathbf{B}_{12} z_2\right) + \underbrace{z_2^{\mathsf{T}} \left(\mathbf{B}_{22} - \mathbf{B}_{21} \mathbf{B}_{11}^{-1} \mathbf{B}_{12}\right) z_2}_{\text{independent of } z_1}.$$

This is the key equation when considering conditional densities. Observing that the term in (8.4) independent on z_1 contributes only to a multiplicative constant, we can write

$$\pi_{Z_1,Z_2}(z_1 \mid z_2) \propto \pi_{Z_1,Z_2}(z_1, z_2)$$

$$\propto \exp\left(-\frac{1}{2}(z_1 + B_{11}^{-1}B_{12}z_2)^T B_{11}(z_1 + B_{11}^{-1}B_{12}z_2)\right).$$

We therefore conclude that the conditional density is Gaussian,

$$\pi_{Z_1 \mid Z_2}(z_1 \mid z_2) \propto \exp\left(-\frac{1}{2}(z_1 - \bar{z}_1)^T D^{-1}(z_1 - \bar{z}_1)\right),$$

with mean

$$\bar{z}_1 = -B_{11}^{-1}B_{12}z_2$$

and covariance matrix

$$D = B_{11}^{-1}.$$

To express these quantities in terms of the covariance matrix C, we need to introduce *Schur complements*. Let us consider the symmetric positive definite matrix $C \in \mathbb{R}^{N \times N}$, partitioned according to (8.2). Since C is positive definite, so are C_{11} and C_{22}. In fact, for any $v_1 \in \mathbb{R}^n$, $v_1 \neq 0$,

$$v_1^T C_{11} v_1 = \begin{bmatrix} v_1^T & 0 \end{bmatrix} \begin{bmatrix} C_{11} & C_{12} \\ C_{21} & C_{22} \end{bmatrix} \begin{bmatrix} v_1 \\ 0 \end{bmatrix} > 0,$$

showing the positive definiteness of C_{11}. The proof that C_{22} is positive definite is analogous.

To compute C^{-1} using the block partitioning, we solve the linear system

$$Cz = y$$

in block form using Gaussian elimination. We begin by partitioning z and y as

$$z = \begin{bmatrix} z_1 \\ z_2 \end{bmatrix} \begin{matrix} \in \mathbb{R}^n \\ \in \mathbb{R}^m \end{matrix}, \quad y = \begin{bmatrix} y_1 \\ y_2 \end{bmatrix} \begin{matrix} \in \mathbb{R}^n \\ \in \mathbb{R}^m \end{matrix},$$

and perform block-wise multiplication to get

$$C_{11}z_1 + C_{12}z_2 = y_1, \tag{8.5}$$

$$C_{21}z_1 + C_{22}z_2 = y_2. \tag{8.6}$$

Solving the second equation for z_2, which can be done because C_{22} is positive definite, thus invertible, we have

$$z_2 = C_{22}^{-1}(y_2 - C_{21}z_1).$$

Substituting this expression for z_2 into the first equation and rearranging the terms yields

$$(C_{11} - C_{12}C_{22}^{-1}C_{21})z_1 = y_1 - C_{12}C_{22}^{-1}y_2.$$

We define the Schur complement of C_{22} to be

$$\tilde{C}_{22} = C_{11} - C_{12}C_{22}^{-1}C_{21}.$$

With this notation,

$$z_1 = \tilde{C}_{22}^{-1}y_1 - \tilde{C}_{22}^{-1}C_{12}C_{22}^{-1}y_2. \tag{8.7}$$

Similarly, solving (8.5) for z_1 first and plugging it into (8.6), we may express z_2 as

$$z_2 = \tilde{C}_{11}^{-1}y_2 - \tilde{C}_{11}^{-1}C_{21}C_{11}^{-1}y_1 \tag{8.8}$$

in terms of the Schur complement of C_{11},

$$\tilde{C}_{11} = C_{22} - C_{21}C_{11}^{-1}C_{12}.$$

Collecting (8.7) and (8.8) into a matrix expression as

$$\begin{bmatrix} z_1 \\ z_2 \end{bmatrix} = \begin{bmatrix} \tilde{C}_{22}^{-1} & -\tilde{C}_{22}^{-1}C_{12}C_{22}^{-1} \\ -\tilde{C}_{11}^{-1}C_{21}C_{11}^{-1} & \tilde{C}_{11}^{-1} \end{bmatrix} \begin{bmatrix} y_1 \\ y_2 \end{bmatrix},$$

we deduce that

$$C^{-1} = \begin{bmatrix} \tilde{C}_{22}^{-1} & -\tilde{C}_{22}^{-1}C_{12}C_{22}^{-1} \\ -\tilde{C}_{11}^{-1}C_{21}C_{11}^{-1} & \tilde{C}_{11}^{-1} \end{bmatrix} = \begin{bmatrix} B_{11} & B_{12} \\ B_{21} & B_{22} \end{bmatrix},$$

which is the formula that we were looking for.

In summary, we have derived the following result.

The conditional density $\pi_{Z_1|Z_2}(z_1 \mid z_2)$ is a Gaussian probability distribution

$$\pi_{Z_1|Z_2}(z_1 \mid z_2) = \mathcal{N}(z_1 \mid \bar{z}_1, D),$$

with conditional mean (CM)

$$\bar{z}_1 = -B_{11}^{-1}B_{12}z_2 = C_{12}C_{22}^{-1}z_2,$$

and conditional covariance

$$D = B_{11}^{-1} = \tilde{C}_{22} = C_{11} - C_{12}C_{22}^{-1}C_{21}.$$

Before showing how these results can be used in some applications, the following observation is in order. If Z_1 and Z_2 are uncorrelated, that is, $C_{12} = 0$, $C_{21} = 0$, the conditional density of Z_1 given Z_2 reduces to the marginal density

$$\pi_{Z_1|Z_2}(z_1 \mid z_2) = \pi_{Z_1}(z_1) = \mathcal{N}(z_1 \mid 0, \mathsf{C}_{11}).$$

In particular, this means that uncorrelated Gaussian random variables are also independent. While vanishing correlation does not, in general, imply independency, in the case of Gaussian random variables it does, and the two concepts are equivalent.

8.2 Linear Inverse Problems

Consider the linear inverse problem

$$b = \mathsf{A}x + e,$$

where $x \in \mathbb{R}^n$ represents the unknown, $b \in \mathbb{R}^m$ is the observed quantity, $e \in \mathbb{R}^m$ is the unknown additive noise, and $\mathsf{A} \in \mathbb{R}^{m \times n}$ is the known forward mapping, and its stochastic extension is

$$B = \mathsf{A}X + E.$$

Assume that X and E are both zero mean Gaussian random variables, and for simplicity, mutually independent,

$$X \sim \mathcal{N}(0, \Gamma), \quad E \sim \mathcal{N}(0, \Sigma),$$

where $\Gamma \in \mathbb{R}^{n \times n}$ and $\Sigma \in \mathbb{R}^{m \times m}$ are symmetric positive definite matrices. We combine X and B into the composite random variable

$$Z = \begin{bmatrix} X \\ B \end{bmatrix} \in \mathbb{R}^{n+m}.$$

In order to apply the formulas of the previous section, we compute the covariance matric C of this random variable. Recalling that

$$\mathrm{E}(XX^{\mathsf{T}}) = \Gamma,$$

and using the independency of X and E and the fact that they have zero means,

$$\begin{aligned}
\mathrm{E}(XB^{\mathsf{T}}) &= \mathrm{E}(X(\mathsf{A}X + E)^{\mathsf{T}}) \\
&= \mathrm{E}(XX^{\mathsf{T}})\mathsf{A}^{\mathsf{T}} + \mathrm{E}(X)\mathrm{E}(E^{\mathsf{T}}) \\
&= \Gamma\mathsf{A}^{\mathsf{T}} \\
&= \mathrm{E}(BX^{\mathsf{T}})^{\mathsf{T}},
\end{aligned}$$

and, similarly,

$$
\begin{aligned}
\mathrm{E}(BB^{\mathsf{T}}) &= \mathrm{E}\big((AX + E)(AX + E)^{\mathsf{T}}\big) \\
&= A\mathrm{E}(XX^{\mathsf{T}})A^{\mathsf{T}} + \mathrm{E}(EE^{\mathsf{T}}) \\
&= A\Gamma A^{\mathsf{T}} + \Sigma.
\end{aligned}
$$

The joint covariance matrix is therefore given by

$$
C = \begin{bmatrix} C_{11} & C_{12} \\ C_{21} & C_{22} \end{bmatrix} = \begin{bmatrix} \Gamma & \Gamma A^{\mathsf{T}} \\ A\Gamma & A\Gamma A^{\mathsf{T}} + \Sigma \end{bmatrix}.
$$

The general result for conditional probability densities of Gaussian variables gives us now the following result.

Given a linear observation model, and assuming a zero mean Gaussian prior and zero mean independent Gaussian additive noise, the posterior density of X given the observation $B = b$ is

$$
\pi_{X,B}(x \mid b) = \mathcal{N}(x \mid \overline{x}, \mathsf{D}),
$$

where the posterior mean and posterior covariance are

$$
\overline{x} = \Gamma A^{\mathsf{T}}(A\Gamma A^{\mathsf{T}} + \Sigma)^{-1}b, \tag{8.9}
$$

$$
\mathsf{D} = \Gamma - \Gamma A^{\mathsf{T}}(A\Gamma A^{\mathsf{T}} + \Sigma)^{-1}A\Gamma. \tag{8.10}
$$

Before considering other applications, an observation about the posterior covariance is in order. Given a direction vector $v \in \mathbb{R}^n$, $\|v\| = 1$, the posterior variance of X along v is given by $v^{\mathsf{T}}\mathsf{D}v$. It follows from (8.10) that

$$
v^{\mathsf{T}}\mathsf{D}v = v^{\mathsf{T}}\Gamma v - v^{\mathsf{T}}\Gamma A^{\mathsf{T}}(A\Gamma A^{\mathsf{T}} + \Sigma)^{-1}A\Gamma v \le v^{\mathsf{T}}\Gamma v,
$$

that is, an indirect observation of a random variable can only decrease the uncertainty about its value.

We remark further that if the prior is not zero mean but rather

$$
X \sim \mathcal{N}(\mu, \Gamma),
$$

the posterior mean needs to be modified as

$$
\overline{x} = \mu + \Gamma A^{\mathsf{T}}(A\Gamma A^{\mathsf{T}} + \Sigma)^{-1}(b - A\mu), \tag{8.11}
$$

while the posterior covariance remains unchanged.

8.3 Interpolation and Conditional Densities

A natural application of the computation of conditional densities for Gaussian distributions is a classic interpolation problem, known in geostatistics as *kriging*, and in probability theory as the Wiener–Kolmogorov prediction.[1] The problem that we consider is stated as follows.

Estimate a smooth function f over a domain $\Omega \in \mathbb{R}^n$ based on few noisy observations of it,

$$b_j = f(t_j) + e_j, \quad t_j \in \Omega, \quad 1 \le j \le m,$$

where e_j represents the additive noise. In geostatistics, where $n = 2$, f could represent the ore distribution over an area, or underwater hydraulic head.

We start by considering the problem in one dimension over a finite interval, $\Omega = [0, 1]$. Let $0 < t_1 < \ldots < t_m < 1$, and assume at first that the values of f at the left and right endpoints are given,

$$f(0) = \beta_L, \quad f(1) = \beta_R.$$

A classic solution to this problem uses cubic splines, i.e., curves that between the interpolation points are third-order polynomials, glued together at the observation points so that the resulting piecewise defined function and its derivatives, up to the second order, are continuous. The Bayesian solution that we are proposing is based on conditioning and, it turns out, can be orchestrated so that the realization with highest posterior probability corresponds to a spline.

Since the statement of the problem does not include an explicit definition of smoothness or of error bounds, we will provide a subjective interpretation. We start by discretizing the problem. For simplicity, we introduce a uniform discretization grid and we assume that the observation points coincide with some of the nodes of the discretization. Let

$$x_k = f(s_k), \quad s_k = \frac{k}{n+1}, \quad 0 \le k \le n+1,$$

for some n, and assume that $t_j = s_{k_j}$ for some k_j, that is, the data are defined as

$$b_j = x_{k_j} + e_j, \quad 1 \le j \le m.$$

We may write the observation model in terms of matrices and vectors. Let $x \in \mathbb{R}^n$ be the vector containing the unknown function values at the interior points, $x_1, \ldots, x_n$, and write the observation model as

$$b = Ax + e,$$

---

[1] The word kriging refers to the name of the South African geostatistician Danie Gerhardus Krige. The other name of the method probably needs no explanations.

where $\mathsf{A} \in \mathbb{R}^{m \times n}$ is the sampling matrix with entries

$$a_{jk} = 1 \text{ if and only if } k = k_j, \ 1 \le j \le m$$
$$a_{jk} = 0 \text{ otherwise.}$$

To account for the smoothness of f, we introduce the second-order smoothness prior

$$2X_1 = \beta_L + X_2 + \gamma W_1$$
$$2X_2 = X_1 + X_3 + \gamma W_2$$
$$\vdots \quad \vdots$$
$$2X_n = X_{n-1} + \beta_R + \gamma W_n,$$

where the random variables W_j are the components of the random vector $W \sim \mathcal{N}(0, \mathsf{I}_n)$. In matrix form, the prior model is given by

$$\mathsf{L}X = \gamma W + \beta,$$

where the vector β accounts for the boundary values,

$$\beta = \begin{bmatrix} \beta_L \\ 0 \\ \vdots \\ 0 \\ \beta_R \end{bmatrix} \in \mathbb{R}^n,$$

and $\mathsf{L} \in \mathbb{R}^{n \times n}$ is the second-order finite difference matrix

$$\mathsf{L} = \begin{bmatrix} 2 & -1 & & \\ -1 & 2 & -1 & \\ & \ddots & \ddots & \ddots \\ & & -1 & 2 \end{bmatrix} \in \mathbb{R}^{n \times n}. \tag{8.12}$$

The matrix L is invertible and, moreover, it allows a factorization in terms of first-order finite difference matrices,

$$\mathsf{L} = \mathsf{L}_1 \mathsf{L}_1^{\mathsf{T}}, \quad \mathsf{L}_1 = \begin{bmatrix} 1 & -1 & & & \\ & 1 & -1 & & \\ & & \ddots & \ddots & \\ & & & & -1 \\ & & & & 1 \end{bmatrix},$$

hence we may write

$$X = \gamma L^{-1} W + L^{-1} \beta,$$

from which we conclude that X is Gaussian, and its prior density is

$$\pi_X(x) = \mathcal{N}(x \mid L^{-1}\beta, \gamma^2 L^{-1} L^{-T}).$$

To model the observation error, we assume that the noise arises from a model

$$E \sim \mathcal{N}(0, \sigma I_m),$$

that is, the additive error is scaled white noise. Therefore, we may write the posterior mean as

$$\begin{aligned}
\bar{x} &= L^{-1}\beta + \gamma^2 L^{-1} L^{-T} A^T (\gamma^2 A L^{-1} L^{-T} A^T + \sigma^2 I_m)^{-1}(b - A L^{-1}\beta) \\
&= L^{-1}\beta + L^{-1} L^{-T} A^T (A L^{-1} L^{-T} A^T + \alpha^2 I_m)^{-1}(b - A L^{-1}\beta),
\end{aligned}$$

where $\alpha = \sigma/\gamma$, and the posterior covariance as

$$D = \gamma^2 L^{-1} L^{-T} - \gamma^2 L^{-1} L^{-T} A^T (A L^{-1} L^{-T} A^T + \alpha^2 I_m)^{-1} A L^{-T} L^{-1}.$$

These rather cumbersome formulas can be slightly simplified by introducing the matrix

$$M = L^{-T} A^T \in \mathbb{R}^{n \times m},$$

so that

$$\begin{aligned}
\bar{x} &= L^{-1}\big(\beta + M(M^T M + \alpha^2 I_m)^{-1}(b - M^T \beta)\big), \\
D &= \gamma^2 L^{-1}\big(I_n - M(M^T M + \alpha^2 I_m)^{-1} M^T\big) L^{-T}.
\end{aligned}$$

Figure 8.1 shows the results of a computed example. Here, the endpoint values are fixed at $\beta_L = 1$ and $\beta_R = 1.5$, and four data points marked with a black dot are given with error standard deviation $\sigma = 0.1$. The prior parameter is set to $\gamma = 10/(n+1)$, where $n = 50$ is the number of discretization intervals. In the figure, we have plotted the posterior mean $\bar{x}$ as well as the credible envelope with an upper and lower boundary defined by

$$c_j^\pm = \bar{x}_j \pm \sqrt{D_{jj}}, \quad 1 \le j \le n.$$

The interpolation approach can be generalized to several dimensions, however, we discuss before some computational issues that simplify the implementation.

Fig. 8.1 Smooth
interpolation of four data
points and fixed endpoint
values. The solid line is the
posterior mean, surrounded
by the one standard deviation
credible envelope

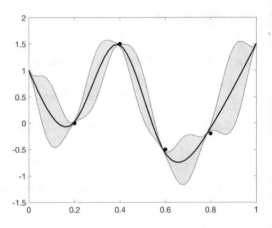

8.4 Covariance or Precision?

In the example discussed in the previous section, the prior covariance matrix was
given by

$$\Gamma = \gamma^2 L^{-1} L^{-T},$$

and its inverse, the precision matrix, is

$$\Gamma^{-1} = \frac{1}{\gamma^2} L^T L.$$

From the computational point of view, in this case the precision matrix is easier to
handle because it is a sparse matrix, while the covariance matrix is a full matrix.
In the example discussed above, the dimensionality of the problem was low enough
($n = 50$) to guarantee that the fullness of the matrix was not a computational issue. In
higher spatial dimensions, however, the sparsity of the matrices used in computations
can make a big difference in terms of computational feasibility. Therefore, it is useful
to consider alternative formulas for the posterior, mean and covariance expressed in
terms of the precision matrix.

Consider the expressions for the probability densities of the Gaussian prior and
Gaussian likelihood,

$$\pi_X(x) \propto \exp\left(-\frac{1}{2} x^T \Gamma^{-1} x\right),$$

$$\pi_{B|X}(b \mid x) \propto \exp\left(-\frac{1}{2}(b - Ax)^T \Sigma^{-1}(b - Ax)\right),$$

and combine them according to Bayes' formula to obtain the probability density of
the posterior,

$$\pi_{X|B}(x \mid b) \propto \pi_X(x)\pi_{B|X}(b \mid x)$$

$$\propto \exp\left(-\frac{1}{2}x^\mathsf{T}\Gamma^{-1}x - \frac{1}{2}(b - \mathsf{A}x)^\mathsf{T}\Sigma^{-1}(b - \mathsf{A}x)\right).$$

Consider the quadratic expression in the exponent of the posterior density,

$$Q(x) = (b - \mathsf{A}x)^\mathsf{T}\Sigma^{-1}(b - \mathsf{A}x) + x^\mathsf{T}\Gamma^{-1}x,$$

and collect the terms of the same order in x to get

$$Q(x) = x^\mathsf{T}(\mathsf{A}^\mathsf{T}\Sigma^{-1}\mathsf{A} + \Gamma^{-1})x - 2x^\mathsf{T}\mathsf{A}^\mathsf{T}\Sigma^{-1}b + b^\mathsf{T}\Sigma^{-1}b$$
$$= x^\mathsf{T}\mathsf{P}x - 2x^\mathsf{T}\mathsf{A}^\mathsf{T}\Sigma^{-1}b + b^\mathsf{T}\Sigma^{-1}b,$$

where

$$\mathsf{P} = \mathsf{A}^\mathsf{T}\Sigma^{-1}\mathsf{A} + \Gamma^{-1}.$$

Completion of the square yields

$$Q(x) = (x - \mathsf{P}^{-1}\mathsf{A}^\mathsf{T}\Sigma^{-1}b)^\mathsf{T}\mathsf{P}(x - \mathsf{P}^{-1}\mathsf{A}^\mathsf{T}\Sigma^{-1}b)$$
$$+ \text{ terms independent of } x,$$

therefore we conclude that the posterior is also Gaussian, and its density is of the form

$$\pi_{X|B}(x \mid b) \propto \exp\left(-\frac{1}{2}(x - \mathsf{P}^{-1}\mathsf{A}^\mathsf{T}\Sigma^{-1}b)^\mathsf{T}\mathsf{P}(x - \mathsf{P}^{-1}\mathsf{A}^\mathsf{T}\Sigma^{-1}b)\right).$$

Therefore the posterior mean and covariance are, respectively,

$$\bar{x} = (\mathsf{A}^\mathsf{T}\Sigma^{-1}\mathsf{A} + \Gamma^{-1})^{-1}\mathsf{A}^\mathsf{T}\Sigma^{-1}b, \tag{8.13}$$
$$\mathsf{D} = (\mathsf{A}^\mathsf{T}\Sigma^{-1}\mathsf{A} + \Gamma^{-1})^{-1}. \tag{8.14}$$

Formula (8.13) for $\bar{x}$ is also known as *Wiener filtered solution*.

It is not immediately obvious that (8.9) and (8.13) are equal, although we know this to be the case because the posterior is uniquely defined, hence the two different derivations of the mean and covariance must agree.

It is possible to prove directly the equality of the two expressions for the mean and for the covariance using linear algebra. The proof, presented below, is based on the *Sherman–Morrison–Woodbury (SMW) identity*, a useful tool to compute efficiently the inverse of the sum of an invertible matrix plus a low rank perturbation. The Sherman–Morrison–Woodbury identity can be formulated as follows.

Let $M \in \mathbb{R}^{n \times n}$ be an invertible matrix, and $U, V \in \mathbb{R}^{n \times k}$, $k \leq n$. Then the matrix

$$M_+ = M + UV^\mathsf{T}$$

is invertible if the $k \times k$ matrix $I_k + V^\mathsf{T} M^{-1} U$ is invertible, and the inverse is given by

$$M_+^{-1} = M^{-1} - M^{-1} U (I_k + V^\mathsf{T} M^{-1} U)^{-1} V^\mathsf{T} M^{-1}. \tag{8.15}$$

Before proving the SMW identity, we show how it can be used to prove the equivalence of the expressions for the mean and the covariance. Consider formula (8.14) for the posterior covariance,

$$D = (A^\mathsf{T} \Sigma^{-1} A + \Gamma^{-1})^{-1}.$$

In the SMW formula, set

$$M = \Gamma^{-1}, \quad U = V = A^\mathsf{T} \Sigma^{-1/2}.$$

Then,

$$I + V^\mathsf{T} M^{-1} U = I + \Sigma^{-1/2} A \Gamma A^\mathsf{T} \Sigma^{-1/2}$$
$$= \Sigma^{-1/2} (\Sigma + A \Gamma A^\mathsf{T}) \Sigma^{-1/2},$$

and, observing that M_+ is the posterior precision, we find that

$$D = M_+^{-1} = \Gamma - \Gamma A^\mathsf{T} (\Sigma + A \Gamma A^\mathsf{T})^{-1} A \Gamma,$$

as claimed.

We are now ready to prove the SMW identity.

Consider the linear equation

$$(M + UV^\mathsf{T}) u = v.$$

First multiply both sides from the left by M^{-1} to obtain

$$u + M^{-1} U V^\mathsf{T} u = M^{-1} v, \tag{8.16}$$

then multiply both sides from the left by V^T to get

$$V^\mathsf{T} u + V^\mathsf{T} M^{-1} U V^\mathsf{T} u = (I_k + V^\mathsf{T} M^{-1} U) V^\mathsf{T} u$$
$$= V^\mathsf{T} M^{-1} v.$$

Solving the last equation for $V^T u$, we obtain

$$V^T u = (I_k + V^T M^{-1} U)^{-1} V^T M^{-1} v. \tag{8.17}$$

On the other hand, from (8.16) we obtain the following expression for u,

$$u = M^{-1} v - M^{-1} U V^T u,$$

and, substituting the right hand side of (8.17) for $V^T u$, gives

$$\begin{aligned} u &= M^{-1} v - M^{-1} U (I_k + V^T M^{-1} U)^{-1} V^T M^{-1} v \\ &= \left(M^{-1} - M^{-1} U (I_k + V^T M^{-1} U)^{-1} V^T M^{-1} \right) v, \end{aligned}$$

which is the alternative formula (8.15) for M_+^{-1}.

8.5 Some Computed Examples

In this section, we consider some computed examples based on the results in this chapter.

8.5.1 Bayesian Interpolation of Multi-Dimensional Data

We return the interpolation problem solved in the case of a one-dimensional signal by kriging. In this case we want to find the interpolant in terms of the alternative formula for the posterior distribution that we have derived in this chapter.

Consider the problem of estimating an unknown two-dimensional smooth function $f : \Omega = [0, 1] \times [0, 1] \to \mathbb{R}$ given a few observed values. We assume here that at the boundary $\partial\Omega$, the function f vanishes. Let the measurements be the values of the function at $y^j = (y_1^j, y_2^j) \in \Omega$, $1 \le j \le m$, and assume that the function values are corrupted by additive scaled Gaussian white noise. We write the stochastic model for the data,

$$B_j = f(y^j) + E_j, \quad 1 \le j \le m,$$

where $E \sim \mathcal{N}(0, \sigma^2 I_m)$. In line with the belief that the underlying function is smooth, we adopt a second-order smoothness prior to X, the random variable comprising the unknown values of f in the $n \times n$ interior points of a regular discretization grid, i.e.,

$$X \sim \mathcal{N}(0, \Gamma),$$

where the covariance matrix is of the form

$$\Gamma = \gamma^2 \Lambda^{-1} \Lambda^{-T}$$

with

$$\Lambda = \mathsf{L}_2 \otimes \mathsf{I}_n + \mathsf{I}_n \otimes \mathsf{L}_2,$$

$\mathsf{L}_2 \in \mathbb{R}^{n \times n}$ being the second-order finite difference matrix, and the parameter γ providing a handle to control the variance of the interpolation. Observe that while the matrix Λ is extremely sparse, thus easy to store and to work with numerically even when the problem is of large dimensions, its inverse is a full matrix. In this case the alternative formulas (8.13) and (8.14) for the posterior mean and covariance turn out to be quite convenient. In fact, in this case the precision matrix, given by

$$\Gamma^{-1} = \frac{1}{\gamma^2} \Lambda^\mathsf{T} \Lambda,$$

is the product of two sparse matrices, and we obtain the following expression for the posterior mean:

$$\overline{x} = \frac{1}{\sigma^2} \left(\frac{1}{\sigma^2} \mathsf{A}^\mathsf{T} \mathsf{A} + \frac{1}{\gamma^2} \Lambda^\mathsf{T} \Lambda \right)^{-1} \mathsf{A}^\mathsf{T} b$$
$$= \left(\mathsf{A}^\mathsf{T} \mathsf{A} + \eta^2 \Lambda^\mathsf{T} \Lambda \right)^{-1} \mathsf{A}^\mathsf{T} b, \quad \eta^2 = \frac{\sigma^2}{\gamma^2}.$$

The computation of the posterior mean requires the solution of the sparse linear system

$$\left(\mathsf{A}^\mathsf{T} \mathsf{A} + \eta^2 \Lambda^\mathsf{T} \Lambda \right) \overline{x} = \mathsf{A}^\mathsf{T} b,$$

which can be done efficiently by means of the iterative solvers described in the next chapter even for very large values of n.

In the computed example, we define a discrete pixel map of size 150×150, and assume that $m = 12$ randomly chosen pixel values are given, and the problem is to smoothly interpolate between these values. The pixel values are assumed to be

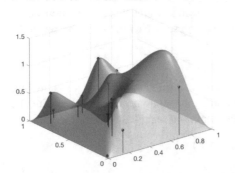 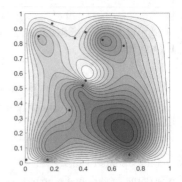

Fig. 8.2 Surface plot (left) and contour plot (right) of the interpolating surface corresponding to the posterior mean with second-order smoothness prior

given with an uncertainty that is expressed in terms of a Gaussian additive error with standard deviation $\sigma = 0.05$. The scaling parameter of the prior was set at $\gamma = 15$ The computed solution is shown in Fig. 8.2.

8.5.2 Posterior Density by Low-Rank Updating

In computational inverse problems, sometimes the data become available in smaller batches, thus it is natural to compute the posterior density by sequential updating steps. Updating strategies using a small subset of the data at a time have been used in some classical inversion algorithms, see Notes and comments. Here, we employ the Sherman–Morrison–Woodbury formula and the two alternative forms for the Gaussian posterior distribution.

Consider a linear inverse problem, written in the stochastic framework as

$$B = AX + E, \quad A \in \mathbb{R}^{m \times n}, \tag{8.18}$$

with Gaussian noise, $E \sim \mathcal{N}(0, \Sigma)$. If

$$\Sigma^{-1} = SS^\mathsf{T}$$

is a symmetric factorization of the precision matrix of the noise, redefining $B = SB$, $A = SA$ we can assume that the noise in (8.18) is white Gaussian noise. By assuming further that X is a priori zero mean Gaussian, we write the model $X \sim \mathcal{N}(0, \Gamma)$.

We start by considering the particular case of the linear problem (8.18) in which $m = 1$, that is, the action of the matrix A can be expressed as an inner product,

$$Ax = u^\mathsf{T}x, \quad u \in \mathbb{R}^n.$$

In this case, with $\Sigma = 1$, formulas (8.13) and (8.14) give for the posterior covariance and mean the formulas

$$D = \left(uu^\mathsf{T} + \Gamma^{-1}\right)^{-1},$$
$$\bar{x} = \left(uu^\mathsf{T} + \Gamma^{-1}\right)^{-1} ub = Dub.$$

We see that the posterior precision matrix is a rank-one update of the prior precision matrix,

$$D^{-1} = uu^\mathsf{T} + \Gamma^{-1},$$

and the update of the covariance matrix can therefore be done using the Sherman–Morrison–Woodbury formula, that is,

$$D = \Gamma - (\Gamma u)\left(1 + u^{\mathsf{T}}(\Gamma u)\right)^{-1} u^{\mathsf{T}}\Gamma$$

$$= \Gamma - \frac{vv^{\mathsf{T}}}{1 + u^{\mathsf{T}}v}, \quad v = \Gamma u,$$

which, too, is a rank-one update. This formula allows us to write the posterior mean as

$$\bar{x} = \Gamma u b - \frac{vv^{\mathsf{T}}}{1 + u^{\mathsf{T}}v} u b$$

$$= b\left(1 - \frac{u^{\mathsf{T}}v}{1 + u^{\mathsf{T}}v}\right)v$$

$$= \frac{b}{1 + u^{\mathsf{T}}v}v.$$

Consider now the general case $A \in \mathbb{R}^{m \times n}$. We write the matrix A in terms of its row vectors as

$$A = \begin{bmatrix} u_1^{\mathsf{T}} \\ \vdots \\ u_m^{\mathsf{T}} \end{bmatrix}, \quad u_j \in \mathbb{R}^n.$$

With this notation, the likelihood density, assuming that the noise has been whitened, can be written as

$$\pi_{B|X}(b \mid x) \propto \exp\left(-\frac{1}{2}\sum_{j=1}^{m}(b_j - u_j^{\mathsf{T}}x)^2\right)$$

$$= \exp\left(-\frac{1}{2}(b_1 - u_1^{\mathsf{T}}x)^2\right)\cdots\exp\left(-\frac{1}{2}(b_m - u_m^{\mathsf{T}}x)^2\right)$$

$$= \prod_{j=1}^{m}\pi_{B_j|X}(b_j \mid x),$$

that is, after whitening, the observations b_j are conditionally independent. We then write the posterior density as

$$\pi_{X|B}(x \mid b) \propto \prod_{j=1}^{m}\pi_{B_j|X}(b_j \mid x)\pi_X(x).$$

For any k, $1 \le k \le m$, we may reinterpret the formula by writing it as

$$\pi_{X|B}(x \mid b) \propto \left(\prod_{j=k+1}^{m} \pi_{B_j|X}(b_j \mid x) \right) \left(\prod_{j=1}^{k} \pi_{B_j|X}(b_j \mid x)\pi_X(x) \right)$$

$$= \left(\prod_{j=k+1}^{m} \pi_{B_j|X}(b_j \mid x) \right) \pi_{X|B_1,\ldots,B_k}(x \mid b_1, \ldots, b_k).$$

This formula suggests a sequential updating of the posterior: Given the prior, we may first compute the posterior $\pi_{X|B_1}(x \mid b_1)$. Setting $k = 1$ in the above formula, we may now consider this posterior as the prior for the remaining observations $B_2, \ldots, B_m$. Inductively, we may then proceed and move one observation at a time from the likelihood to the updated prior, until the full observation vector is processed.

We summarize the idea in algorithmic form.

1. **Initialize:** Set the current mean and covariance $\bar{x} = 0$, Γ. Set the counter $j = 0$.
2. **Iterate:** While $j < m$,

   (a) compute the vector $v_j = \Gamma u_j$,
   (b) update the mean and the covariance,

   $$\bar{x}_+ = \bar{x} + \frac{b_j - u_j^\mathsf{T}\bar{x}}{1 + u_j^\mathsf{T}v_j} v_j,$$

   $$\Gamma_+ = \Gamma - \frac{v_j v_j^\mathsf{T}}{1 + u_j^\mathsf{T}v_j},$$

   (c) replace the current values by the updated ones, $\bar{x} = \bar{x}_+$, $\Gamma = \Gamma_+$.
   (d) Advance the counter, $j = j + 1$.

When completed, $\bar{x}$ and Γ contain the posterior mean and covariance.

We demonstrate the algorithm with a small computed example.

Consider the deconvolution problem of recovering a function $f : [0, 1] \to \mathbb{R}$ from few convolution data,

$$g(s_k) = \int_0^1 a(s_k - t)f(t)dt + e_k$$

$$\approx \frac{1}{n} \sum_{j=1}^{n} a(s_k - t_j)f(t_j) + e_k, \quad t_j = \frac{j}{n},$$

or in matrix notation, we can write the system as (8.18), where the components x_j correspond to the values of f at the interior points t_j, $1 \le j \le n - 1$. We choose the kernel a to be a Gaussian,

$$a(t) = \frac{1}{2}e^{-t^2/(2w^2)}, \quad w = 0.08.$$

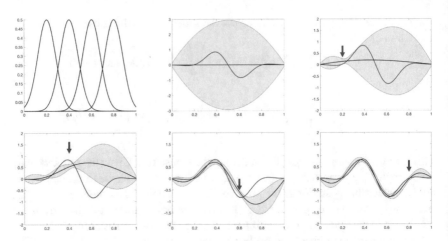

Fig. 8.3 From top left in lexicographical order: The four row vectors of the matrix A corresponding to the convolution kernels. Prior envelope of two marginal standard deviations, and the generative function f^* plotted in red. Updated posterior means (black curves) with the posterior marginal of two standard deviation envelopes after one, two, three and four updates. The red arrow indicates the center point of the convolution kernel at the updating round

We select the data points at $s_1 = 0.2$, $s_2 = 0.4$, $s_3 = 0.6$ and $s_4 = 0.8$, and the discretization level is set at $n = 100$. The four convolution kernels constituting the rows of A are shown in the top left panel of Fig. 8.3. We generate the data using a generative model

$$f^*(t) = (0.5 \sin 2\pi t - 0.7 \sin 4\pi t)\, e^{-15t^2}.$$

Additive independent Gaussian noise with standard deviation $\sigma = 0.001$ is subsequently added to the data, and the noise is whitened. For the prior, we use the second-order smoothness prior with presumably known boundary values $f(0) = f(1) = 0$, that is,

$$\Gamma^{-1} = \gamma^2 \mathsf{L}^\mathsf{T}\mathsf{L}, \quad \mathsf{L} = \begin{bmatrix} -2 & 1 & & \\ 1 & -2 & \ddots & \\ & \ddots & \ddots & 1 \\ & & 1 & -2 \end{bmatrix} \in \mathbb{R}^{(n-1)\times(n-1)}.$$

The parameter γ is chosen so that the estimated marginal variances correspond to our expectations of the size of the unknown function. In the current example, the value $\gamma = 1$ was used. In Fig. 8.3, the second panel in the top row shows the generative function f^*, and the prior marginal envelope of two standard deviations, that is, the shaded area lies between the values

$$\mu_j \pm 2\sqrt{\Gamma_{jj}} = \pm 2\sqrt{\Gamma_{jj}}, \quad 1 \le j \le n - 1,$$

as the prior was assumed to be zero mean.

We run the four updating steps, and after each update plot the posterior mean and the marginal envelopes of two standard deviations. The progression of the iteration is shown in Fig. 8.3. We observe that every time a new update is computed, the posterior variance is reduced significantly near the maximum of the corresponding convolution kernel.

Notes and Comments

Sequential updating algorithms using a low-dimensional portion of the data at a time were particularly popular in earlier times, when computers had rather restricted memory capacity. One of such methods is known as Kaczmarz iteration, or Algebraic Reconstruction Technique (ART) in computerized tomography, where the data are entered one projection at a time [57]. Low-rank updates are also popular in optimization by Newton-type methods, where one needs to update the Hessian as the iterations proceed [32].

For the connection between the spline approximation and Bayesian models, we refer to the article [51].

Chapter 9
Iterative Linear Solvers
and Priorconditioners

In this chapter we return to the solution of linear systems of equations, a task that in scientific computing has a core role, either as a problem of interest per se or as part of a larger computational endeavor. We begin by discussing iterative linear system solvers and how they are used for the solution of ill-conditioned inverse problems following a classical approach.

Traditionally, when solving an ill-conditioned linear system $Ax = b$, e.g., in the least squares sense, the remedies for ill-posedness are selected based on the properties of the matrix A rather than on the expectations about the sought solution x. In contrast, in the Bayesian framework the beliefs about the solution play an active role. After reviewing the classical use of iterative solvers for ill-conditioned linear systems, we will adopt the Bayesian perspective and outline a strategy for importing a priori beliefs in the solvers, taking advantage of the technical tools of numerical linear algebra. In particular, statistically inspired preconditioners expressing the believed properties of the solution are ideally positioned to bridge the deterministic and the Bayesian paradigms for solving linear ill-posed problems.

9.1 Iterative Methods in Linear Algebra

The solution of a linear system by an iterative solver is inherently very different from the solution by a direct solver, because the solution is formed gradually, through a sequence of steps, starting from an initial approximate guess. At each step the way in which the matrix of the coefficients or its transpose enter is through their multiplicative action, thus the matrix itself is not needed in explicit form, but only a protocol for computing the needed matrix-vector products. Iterative methods that do not require access to the matrix in explicit form are called *matrix free*.

In general, iterative methods for computing the solution of a system $b = Ax$, start from an initial guess x_0 and compute a sequence

© The Author(s), under exclusive license to Springer Nature Switzerland AG 2023
D. Calvetti and E. Somersalo, *Bayesian Scientific Computing*, Applied Mathematical
Sciences 215, https://doi.org/10.1007/978-3-031-23824-6_9

$$x_1, x_2, \ldots, x_k, \ldots,$$

of improved approximate solutions. Since the matrix A does not need to be explicitly formed, iterative linear solvers have long been the methods of choice when the matrix A is either not explicitly known, or of very large dimensions relative to the computer memory, while its action on a vector can be computed easily and effectively. The matrix action $x \mapsto Ax$ can be given in the form of an algorithm. Another instance when iterative methods are preferable to direct linear solvers is when it suffices to find an approximate solution of the linear system.

In general, even if an iterative solver is guaranteed to converge to the desired solution, the process is terminated when either a maximum number of iterations has been reached, or some convergence criterion has been satisfied. Iterative linear solvers typically stop when the norm of the error in the approximation of the data,

$$r_k = b - Ax_k,$$

referred to as the *residual error* in the linear algebra literature, or *discrepancy* in the inverse problems literature, has been sufficiently reduced. The approximate solutions determined during the iteration steps are called the *iterates*.

The availability of an approximate solution at any stage of the process is one of the properties that differentiate iterative solvers most from direct methods, where the entire process needs to run to completion before any information about the solution is available. For that reason, direct methods are of the *all or nothing* kind, because either the solution is returned at the end of the process, or all is lost if the process is prematurely interrupted.

The literature on iterative linear solvers is rather vast and comprises different families of iterative schemes. In this book we limit our discussion to a class of Krylov subspace iterative solvers, introduced next, where the iterates belong to a nested sequence of linear subspaces that are automatically determined in the iteration steps. The subspaces are completely determined from the right-hand side vector and the coefficient matrix, thus the iterative schemes are effectively *nonlinear solvers for linear systems*. This fact makes them extremely versatile and efficient, but also more difficult to analyze.

9.2   Krylov Subspace Iterative Methods

We set up the stage for our iterative methods of choice by defining the subspaces that give them their name, and where the iterates live. We start by considering square matrices. Given a matrix $A \in \mathbb{R}^{n \times n}$ and a vector $b \in \mathbb{R}^n$, the kth *Krylov subspace* associated with the pair (A, b) is

$$\mathcal{K}_k(A, b) = \mathrm{span}\{b, Ab, \ldots, A^{k-1}b\}.$$

The two iterative linear solvers that we will introduce below are characterized by the fact that the kth iterate belongs to a suitable kth Krylov subspace, although the matrix-vector pair determining the sequence of Krylov subspaces and the criterion for determining the kth iterate will differ for the two schemes. Iterative schemes whose subsequent iterates are in a nested sequence of Krylov subspaces are called *Krylov subspace iterative methods*.

9.2.1 Conjugate Gradient Algorithm

The first Krylov subspace iterative linear solver, originally proposed by Hestenes and Stiefel in 1952 (see Notes and Comments at the end of the chapter) as a variation on Gram–Schmidt orthogonalization for symmetric positive definite (SPD) matrices, is the *Conjugate Gradient (CG) method*. The CG method was proposed originally as an alternative to Gaussian elimination for the solution of the special class of linear systems with SPD matrices, and the sequence of its iterates was guaranteed to converge to the solution in at most as many steps as the dimension of the matrix. The popularity of the CG method increased tremendously with the wider availability of computers, and modifications of the algorithm for symmetric positive semi-definite matrices were proposed in the literature.

One reason why the CG method is very popular for the solution of large-scale problems is the fact that it requires only one matrix-vector product with the matrix A per iteration, and that its memory allocation is very small and essentially independent of the number of iterations. The kth iterate determined by the CG method implicitly minimizes the A-norm of the error,

$$x_k = \arg \min_{x \in \mathcal{K}_k(A,b)} \|x - x^*\|_A^2,$$

where the A-norm is defined as

$$\|z\|_A^2 = z^T A z,$$

and $x^* = A^{-1}b$ is the *unknown* exact solution that we want to compute. Fortunately, the CG algorithm is a procedure for computing the minimizer x_k without knowing x^*. A detailed derivation of the CG algorithm is beyond the scope of this book and will not be presented, instead we briefly outline the ideas behind it.

At each step $k \geq 1$, the CG algorithm searches for the scalar α which minimizes the functional

$$\alpha \mapsto \|x_{k-1} + \alpha p_{k-1} - x^*\|_A^2,$$

where x_{k-1} is the previous approximate solution and p_{k-1} is a vector, called *search direction* because it is the direction of the correction to the previous iterate. The minimizer, found by setting the derivative of the functional with respect to α equal to zero, is

$$\alpha_k = \frac{\|r_{k-1}\|^2}{p_{k-1}^{\mathsf{T}} \mathsf{A} p_{k-1}},$$

where r_{k-1} is the residual error of the previous approximation,

$$r_{k-1} = b - \mathsf{A} x_{k-1},$$

and the new approximate solution is

$$x_k = x_{k-1} + \alpha_k p_{k-1}.$$

The selection of the search directions is crucial. In the first iteration, the search direction is determined by the residual error associated with the initial guess,

$$p_0 = r_0 = b - \mathsf{A} x_0.$$

In each subsequent iteration, the new search direction p_k is a vector A-*conjugate* to all previous search directions, hence it must satisfy

$$p_k^{\mathsf{T}} \mathsf{A} p_j = 0, \quad 0 \le j \le k - 1.$$

While at first sight this requirement looks quite complicated and its numerical implementation time consuming, it turns out that each new search direction lies in the subspace spanned by the previous search direction and the most recent residual error vector, that is,

$$p_k = r_k + \beta_k p_{k-1}, \quad r_k = b - \mathsf{A} x_k.$$

The parameter β_k is chosen so that the A–conjugacy is satisfied. After some algebraic manipulation it turns out that

$$\beta_k = \frac{\|r_k\|^2}{\|r_{k-1}\|^2}.$$

We are now ready to outline how to organize the computations for the Conjugate Gradient method. For simplicity, the initial guess is set to zero here.

The Conjugate Gradient (CG) algorithm: Given the right-hand side b and an algorithm to multiply a vector by A:

1. **Initialize:**
   $x_0 = 0;$
   $r_0 = b - \mathsf{A} x_0;$
   $p_0 = r_0;$
   Set the counter $k = 1$.

2. **Iterate** until a stopping criterion is met:

$$\alpha = \frac{\|r_{k-1}\|^2}{p_{k-1}^T A p_{k-1}};$$

$$x_k = x_{k-1} + \alpha p_{k-1};$$
$$r_k = r_{k-1} - \alpha A p_{k-1};$$

$$\beta = \frac{\|r_k\|^2}{\|r_{k-1}\|^2};$$

$$p_k = r_k + \beta p_{k-1};$$
Advance the counter $k \to k + 1$.

If the initial guess is the zero vector as we assume here, the first k iterates determined by the CG algorithms live in the nested sequence of subspaces,

$$\text{span}\{b\} \subset \text{span}\{b, Ab\} \subset \cdots \subset \text{span}\{b, Ab, \ldots, A^{k-1}b\},$$

as can be easily verified by the construction of the sequence.

9.2.2 Conjugate Gradient Method for Least Squares

In general, the conjugate gradient method will break down if the matrix A is not symmetric positive definite. If the matrix A is not square and we want to compute the least squares solution of the associated linear system, and if the columns of A are linearly independent, the matrix $A^T A$ of the associated normal equations,

$$A^T A x = A^T b, \tag{9.1}$$

is SPD, and the CG methods can be used for computing the solution. The seminal 1952 paper by Hestenes and Stiefel described a variation of the CG algorithm which can be used for computing the least squares solution of linear systems with general, non-square matrix A. The algorithm, known as the *Conjugate Gradient method for Least Squares* (CGLS) is mathematically equivalent to applying the CG method to the normal equations (9.1) *without ever forming the matrix* $A^T A$. The CGLS method is computationally more expensive than the CG method, requiring two matrix-vector products per iteration, one with A, one with A^T, but its memory allocation is essentially insensitive to the number of iterations. It can be shown that the kth CGLS iterate solves the minimization problem

$$x_k = \arg \min_{x \in \mathcal{K}_k(A^T A, A^T b)} \|b - Ax\|^2,$$

where the associated Krylov subspace is, by definition,

$$\mathcal{K}_k(\mathsf{A}^\mathsf{T}\mathsf{A}, \mathsf{A}^\mathsf{T}b) = \mathrm{span}\{\mathsf{A}^\mathsf{T}b, (\mathsf{A}^\mathsf{T}\mathsf{A})\mathsf{A}^\mathsf{T}b, \dots, (\mathsf{A}^\mathsf{T}\mathsf{A})^{k-1}\mathsf{A}^\mathsf{T}b\}.$$

Equivalently, the kth iterate is characterized by

$$\Phi(x_k) = \min_{x \in \mathcal{K}_k(\mathsf{A}^\mathsf{T}\mathsf{A}, \mathsf{A}^\mathsf{T}b)} \Phi(x),$$

where

$$\Phi(x) = \frac{1}{2}x^\mathsf{T}\mathsf{A}^\mathsf{T}\mathsf{A}x - x^\mathsf{T}\mathsf{A}^\mathsf{T}b.$$

The strategy for determining the minimizer is very similar to that of the Conjugate Gradient method. The search directions,

$$p_0, p_1, \dots, p_{k-1}, p_k, \dots,$$

determined in the iterations are $\mathsf{A}^\mathsf{T}\mathsf{A}$-*conjugate directions*, hence, by definition, the vectors p_j satisfy the condition

$$p_j^\mathsf{T}\mathsf{A}^\mathsf{T}\mathsf{A}p_k = 0, \quad 0 \le j \le k-1..$$

The iterate x_k is determined by updating x_{k-1} according to the formula

$$x_k = x_{k-1} + \alpha_k p_{k-1},$$

where the coefficient $\alpha_k \in \mathbb{R}$ solves the minimization problem

$$\alpha_k = \arg\min_{\alpha \in \mathbb{R}} \Phi(x_{k-1} + \alpha p_{k-1}).$$

Introducing the *residual error of the normal equations* for the CGLS method associated with x_k,

$$r_k = \mathsf{A}^\mathsf{T}b - \mathsf{A}^\mathsf{T}\mathsf{A}x_k,$$

the passage from the current search direction to the next is given by

$$p_k = r_k + \beta_k p_{k-1},$$

where the coefficient β_k is chosen so that p_k is $\mathsf{A}^\mathsf{T}\mathsf{A}$-conjugate to the previous search directions. It can be shown that

$$\beta_k = \frac{\|r_k\|^2}{\|r_{k-1}\|^2}.$$

The quantity

$$d_k = b - Ax_k$$

is called the *discrepancy* associated with x_k. The discrepancy and the residual of the normal equations are related via the equation

$$r_k = A^T d_k.$$

It can be shown that the norms of the discrepancies form a non-increasing sequence, while the norms of the solutions form a non-decreasing one,

$$\|d_{k+1}\| \leq \|d_k\|, \qquad \|x_{k+1}\| \geq \|x_k\|.$$

We are now ready to describe how the calculations for the CGLS method should be organized. For simplicity, the initial guess is set to zero.

The Conjugate Gradient for Least Squares (CGLS) algorithm: Given the right-hand side b, and algorithms to multiply by A and A^T.

1. **Initialize:**
   $x_0 = 0;$
   $d_0 = b - Ax_0;$
   $r_0 = A^T d_0;$
   $p_0 = r_0;$
   $y_0 = Ap_0;$
   Set the counter $k = 1$.

2. **Iterate** until the stopping criterion is met:
   $$\alpha = \frac{\|r_{k-1}\|^2}{\|y_{k-1}\|^2}$$

   $x_k = x_{k-1} + \alpha p_{k-1};$
   $d_k = d_{k-1} - \alpha y_{k-1};$
   $r_k = A^T d_k;$

   $$\beta = \frac{\|r_k\|^2}{\|r_{k-1}\|^2};$$

   $p_k = r_k + \beta p_{k-1};$
   $y_k = Ap_k;$
   Advance the counter $k \rightarrow k + 1$.

Observe that when the columns of the matrix $A \in \mathbb{R}^{m \times n}$ are linearly independent, the CGLS algorithm converges to the minimum norm solution of the least squares problem in at most m iterations. Furthermore, if A is invertible, the minimum norm solution is also the solution to the equation $Ax = b$.

The CGLS algorithm, combined with Bayesian prior models, turns out to be an extremely useful tool for solving inverse problems. Observe that we have not specified the stopping criterion for the CG or the CGLS algorithm, because, as it turns out, the stopping criterion for inverse problems plays an important role that will be discussed more in detail in the following section.

9.3   Ill-Conditioning and Errors in the Data

In real applications, the discretization of a linear relation linking the unknown to be determined and the observed data gives rise to linear systems which can be moderately to severely ill-conditioned. The condition number of the matrix representation of the discretized forward model is usually regarded as an indicator of how severely error corrupting the data may be amplified in the process of solving the linear system. If the linear system is solved by a direct method, there is no way to monitor the growth of the amplified error, while when using iterative methods the contribution of the error components accumulates with the number of steps. In fact, often at first the iterates seem to converge to a meaningful solution, then, as the iterations proceed, they begin to diverge, a phenomenon known as *semi-convergence*.

Example 9.1 Consider the problem of numerical differentiation. We start with a differentiable function $f : [0, 1] \rightarrow \mathbb{R}$, satisfying the boundary condition $f(0) = 0$, and assume that discrete noisy observations of the function values constitute our data,

$$b_j = f(t_j) + e_j, \quad t_j = \frac{j}{n}, \quad 1 \leq j \leq n.$$

The problem is to estimate the derivative of f at points t_j, that is, the unknowns are given by

$$x_j = f'(t_j), \quad 1 \leq j \leq n.$$

In our computed example, we consider a function

$$f(t) = 1 + \mathrm{erf}\left(6\left(t - \frac{1}{2}\right)\right), \tag{9.2}$$

where erf is defined as

$$\mathrm{erf}(t) = \frac{2}{\sqrt{\pi}} \int_0^t e^{-s^2} ds,$$

the sigmoidal error function, and we have

$$f'(t) = \frac{12}{\sqrt{\pi}} e^{-36(t-1/2)^2}.$$

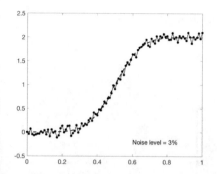
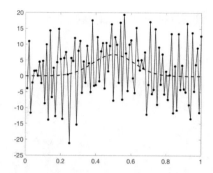

Fig. 9.1 Left: Numerical differentiation data. The noiseless function is plotted with dashed curve. Right: Numerical derivative by naïve finite differencing. The true solution is indicated by the dashed curve

We set $n = 100$, and add scaled white noise to the discretized data, with a standard deviation 3% of the maximum of the noiseless signal. The noiseless function f and the noisy data are shown in Fig. 9.1.

A naïve and straightforward solution would try to use finite differences, writing

$$x_j \approx \frac{f(t_j) - f(t_{j-1})}{h}, \quad h = \frac{1}{n},$$

and simply to substitute the noisy approximations $b_j \approx f(t_j)$ into this formula. The computed solution is rather meaningless, as shown in Fig. 9.1. It is also rather easy to understand why: Substitution of the noisy data into the finite difference formula gives

$$\frac{b_j - b_{j-1}}{h} = \frac{f(t_j) - f(t_{j-1})}{h} + \frac{e_j - e_{j-1}}{h} \approx f'(t_j) + n(e_j - e_{j-1}),$$

that is, the finite difference formula for the data produces the correct approximation with an additive error term amplified by a factor n which completely overwhelms the approximation.

To set the problem in linear algebraic terms, we first observe that, by the fundamental theorem of calculus,

$$f(t_j) = \int_0^{t_j} f'(t)dt = \sum_{k=1}^{j} \int_{t_{k-1}}^{t_k} f'(t)dt,$$

which we approximate as

$$f(t_j) \approx h \sum_{k=1}^{j} x_k.$$

Thus, the problem allows a discrete approximate representation

$$b = \mathsf{A}x + e,$$

where $\mathsf{A} \in \mathbb{R}^{n \times n}$ is given by

$$\mathsf{A} = h \begin{bmatrix} 1 & & & \\ 1 & 1 & & \\ \vdots & & \ddots & \\ 1 & 1 & \cdots & 1 \end{bmatrix} \in \mathbb{R}^{n \times n}.$$

Observe that the CGLS algorithm does not require that we form the matrix A explicitly, as long as we have an algorithm to compute its action on a vector. In this case, the matrix-vector product has components

$$(\mathsf{A}x)_j = h \sum_{k=1}^{j} x_k,$$

a cumulative sum of the entries of x, which in Matlab can be effectuated by the command cumsum. Similarly, the transpose of A is an upper triangular matrix of ones, and the algorithmic definition of its action is

$$\left(\mathsf{A}^\mathsf{T} y\right)_j = h \sum_{k=j}^{n} y_k.$$

We implement the CGLS algorithm for this problem and verify the semi-convergence phenomenon. Figure 9.2 shows three plots, from left to right: The norm of the discrepancy, $\|b - \mathsf{A}x_j\|$ as a function of the iteration count j, the norm of the error of the approximation x_j with respect to the true value,

$$\delta_j = \|x_j - x^*\|,$$

where x^* is the point value of f given by (9.2), and finally, the norm $\|x_j\|$ of the approximation. Observe that in realistic problems, we don't have the true solution, so the quantity δ_j is not available. As expected, the discrepancy decreases monotonically, while the norm of the solution keeps increasing. The norm of the error, however, demonstrates the semi-convergence behavior: At the beginning, the approximation error keeps decreasing, but at some point, the approximation starts to deteriorate, indicating that the algorithm starts to fit the approximation to the noise that soon takes over.

Finally, in Fig. 9.3 we show the first eight iterates computed by CGLS.

The previous example indicates that when using the CGLS method to solve a linear discrete inverse problem with a noisy right-hand side, it is important to stop

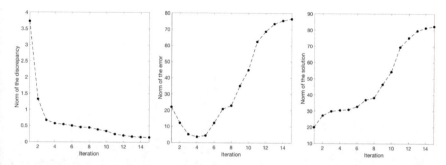

Fig. 9.2 The norm of the discrepancy as a function of the iteration, $\|b - Ax_j\|$ (left), the norm of the error of the approximation, $\|x_j - x^*\|$ (center), and the norm of the approximation, $\|x_j\|$ (right)

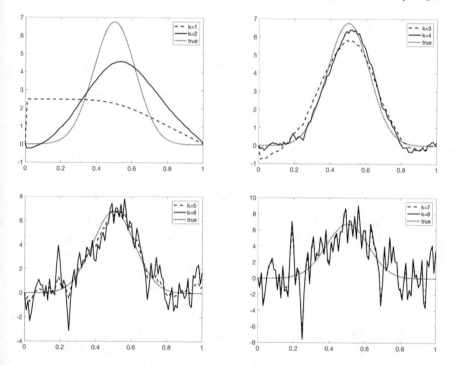

Fig. 9.3 The approximate solutions computed by CGLS iteration, in lexicographical order from iteration 1 to iteration 8. After the eighth iteration, the amplitude of the noise in the reconstruction dominates

the iteration before the amplified error begins to dominate the computed solution. To achieve this, the CGLS algorithm should be equipped with a stopping rule preventing amplified noise components from being included in the solution, thus making the solution less sensitive to noise in the data. The termination of the iterations before the reduction of the residual error has been completed is effectively a way of regularizing the problem, known as *regularization by early stopping*. The design of the stopping

Fig. 9.4 The residual norm
as a function of the iteration,
and the norm of the noise

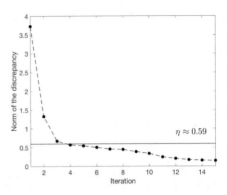

rule for an iterative solver of linear discrete inverse problems is an essential part of
the approximate solver. We outline here how this problem can be approached in a
fully deterministic way, prior to revisiting the issue from a Bayesian perspective.

Assume for a moment that the data vector b represents a noisy observation of a
vector b^* that presumably corresponds to an underlying exact solution $x^* \in \mathbb{R}^n$, that
is,

$$b = b^* + e = Ax^* + e,$$

and that, while the additive noise is unknown, an estimate of its size is available, i.e.,

$$\|e\| \approx \eta, \quad \eta > 0 \text{ known.}$$

Thus, Ax^* is a vector that lies roughly at the distance η from the data, but in an
unknown direction, suggesting that any vector x for which $\|Ax - b\| \le \eta$ should be
as good as any other approximation. To safeguard against underestimating the norm
of the noise, the condition for acceptable solutions is often stated as

$$\|Ax - b\| \le \tau\eta, \tag{9.3}$$

where the parameter $\tau > 1$ is a safeguard factor whose value is close to one.

The idea of using (9.3) to decide how much a problem should be regularized is
usually referred to as the *Morozov discrepancy principle*. When applied to iterative
solvers, the Morozov discrepancy principle stipulates that the iterations should be
stopped at the iteration j^* satisfying

$$\|Ax_{j^*-1} - b\| > \tau\eta \ge \|Ax_{j^*} - b\|.$$

Observe that since the norms of the discrepancies for CGLS form a non-increasing
sequence, the stopping criterion is unambiguous, assuming that the required value
for the norm of the discrepancy is eventually reached.

In Fig. 9.4, we have plotted the residual norm of the previous example and indicated the norm of the additive noise. The plot indicates that Morozov's discrepancy principle with $\tau = 1$ would suggest the fourth iteration to be a suitable solution, which in the light of the semi-convergence analysis is also the best solution. Observe, however, that in this synthetic case, we were able to evaluate the norm of the noise vector, while in real applications, the norm can be at best a rough estimate based on the statistical properties of the noise, as we shall see when moving to the statistical version of the principle.

9.4 Iterative Solvers in the Bayesian Framework

The solution of a linear system by Krylov subspace iterative solvers is sought in a subspace that depends on the data and the matrix defining the system. If we want to incorporate prior information about the solution into the subspaces, we need to modify the linear system using the prior information. How to do this in a computationally efficient manner is the topic of this section.

We start by briefly reviewing some of the basic facts and results about preconditioners which we will be using later. Given a linear system of equations

$$Ax = b, \quad A \in \mathbb{R}^{n \times n}, \tag{9.4}$$

with A invertible, and a nonsingular matrix $M \in \mathbb{R}^{n \times n}$, it is immediate to verify that the linear system

$$M^{-1}Ax = M^{-1}b \tag{9.5}$$

has the same solution as (9.4). Likewise, if $R \in \mathbb{R}^{n \times n}$ is a nonsingular matrix, the linear system

$$AR^{-1}w = b, \quad Rx = w, \tag{9.6}$$

has also the same solution x as (9.4).

The *convergence rate* of an iterative method for the solution of a linear system depends typically on the spectral properties of the matrix A. Thus, if we replace the linear system (9.4) with (9.5) or (9.6), the rate of convergence will depend on the spectral properties of $M^{-1}A$ or AR^{-1}, respectively, instead of on the spectral properties of A. It is therefore clear that if M or R are suitably chosen, the convergence of an iterative method for (9.5) or (9.6), respectively, may be much faster than for (9.4).

The matrix M (R, respectively) is called a *left (right) preconditioner* and the linear system (9.5) (or (9.6)) is referred to as the *left (right) preconditioned linear system*.

Naturally, it is possible to combine both right and left preconditioners and consider the system

$$\mathsf{M}^{-1}\mathsf{A}\mathsf{R}^{-1}w = \mathsf{M}^{-1}b, \quad \mathsf{R}x = w.$$

In the statistical discussion to ensue, the left and right preconditioners will have different roles: one will be related to the likelihood, the other to the prior. Moreover, we are less interested in the convergence properties than in the qualities of the solution, as expressed implicitly by the prior.

9.4.1 Preconditioning and Tikhonov Regularization

In general, the closer the matrix A is to the identity, the easier it becomes for a Krylov subspace iterative method to sufficiently reduce the norm of the residual error. Thus, in the traditional linear algebraic framework, for the construction of good preconditioners it is desirable to find a matrix M or R such that $\mathsf{A}\mathsf{R}^{-1}$ or $\mathsf{M}^{-1}\mathsf{A}$ is close to the identity.

Since the application of an iterative method to a preconditioned linear system will require the computation of matrix-vector products of the form $\mathsf{M}^{-1}\mathsf{A}x$ or $\mathsf{A}\mathsf{R}^{-1}w$, followed by the solution of linear systems of the form $\mathsf{R}x = w$ in the case of right preconditioners, it is also important that these computations can be done efficiently.

In the general case, the question of whether we should use left or right preconditioning depends only on the iterative method that we are using, and on whether or not there are reasons to keep the right-hand side of the original linear system unchanged. Furthermore, the best general purpose preconditioners are those which yield fastest rate of convergence for the least amount of work.

The situation becomes a little different when dealing with linear systems of equations with an ill-conditioned matrix and a right-hand side contaminated by noise. We have seen in the previous sections that as the iteration number increases, amplified noise components start to corrupt the computed solution. Ideally, a good preconditioner speeds up the convergence of the iterations without speeding up the rate at which the amplified noise takes over the solution. The sensitivity of a linear system to noise is related to the smallest eigenvalues of the matrix, or more generally, its smallest singular values. In an effort to accelerate the rate of convergence of the iterative methods, while keeping the noise in the right-hand side from overtaking the computed solution, preconditioners that enhance the convergence in the subspace corresponding to singular vectors associated with large singular values of A have been proposed in the literature. This separation may be difficult if there is no obvious gap in the singular values of A. Furthermore, the separation should also depend on the noise level. Moreover, finding the spectral information of the matrix may be computationally challenging and costly.

In the Bayesian context, the main interest is in the properties of the unknown, some of which we may have expressed in the form of a priori belief, and the preconditioner is the Trojan horse carrying this belief into the algorithm. Thus, instead of getting our inspiration about the selection of preconditioners from the general theory about iterative system solvers, we examine them in the light of linear Gaussian models that are closely related also to Tikhonov regularization.

The design of regularizing preconditioners starts traditionally from Tikhonov regularization, which is a bridge between statistical and deterministic solution of inverse problems. Recall that instead of seeking a solution to the ill-conditioned linear system $Ax = b$, the strategy in Tikhonov regularization is to replace the original problem by a nearby minimization problem,

$$x_\alpha = \arg\min \left(\|Ax - b\|^2 + \alpha \|Rx\|^2 \right), \tag{9.7}$$

where the penalty term is selected in such a way that for the desired solution, the norm of Rx is not excessively large, and the regularization parameter α is chosen, e.g., by the Morozov discrepancy principle.

The simplest version of Tikhonov regularization method—sometimes, erroneously, called *the* Tikhonov regularization method, while calling (9.7) a generalized Tikhonov regularization (see Notes and Comments of Chap. 7)[1]—is to choose $R = I_n$, the identity matrix, leading to the minimization problem

$$x_\alpha = \arg\min \left(\|Ax - b\|^2 + \alpha \|x\|^2 \right). \tag{9.8}$$

Recalling the argument in favor of early termination of iterative solvers to avoid the norm of the solution to become excessively large, and observing that avoiding such increase is exactly the reason for introducing the penalty term in (9.8), we conclude that a suitable alternative to Tikhonov regularization (9.8) is to use iterative solvers with an early stopping rule. In particular, when the CGLS method is employed, the norms of the iterates form a non-decreasing sequence, yielding a straightforward way of monitoring the growth of the penalty term $\|x\|$.

So what could be an alternative to the more general regularization strategy (9.7)? It is obvious that if R is a nonsingular square matrix, the regularization strategy (9.7) is equivalent to

$$w_\alpha = \arg\min \left(\|AR^{-1}w - b\|^2 + \alpha \|w\|^2 \right), \quad Rx_\alpha = w_\alpha,$$

and the natural candidate for an alternative strategy to Tikhonov regularization would be a iterative solver with early stopping and right preconditioner R.

To understand the effects of choosing preconditioners in this fashion, we go back to the Bayesian theory, which was our starting point to arrive at Tikhonov regularization. We consider the linear additive model

$$B = AX + E, \quad E \sim \mathcal{N}(0, \sigma^2 I), \quad X \sim \mathcal{N}(0, \Gamma),$$

---

[1] Andrei Nikolaevich Tikhonov (1906–1993), who had a great impact on various areas of mathematics, was originally a topologist, and, loyal to his background, thought of regularization in terms of compact embeddings.

leading to the posterior probability density

$$\pi_{X|B}(x \mid b) \propto \exp\left(-\frac{1}{2\sigma^2}\|b - Ax\|^2 - \frac{1}{2}x^{\mathsf{T}}\Gamma^{-1}x\right).$$

Consider the prior covariance in terms of its eigenvalues and eigenvectors. Let

$$\Gamma = W\Lambda W^{\mathsf{T}}$$

be the eigenvalue decomposition of the covariance matrix Γ, where W is an orthogonal matrix whose columns $w_j \in \mathbb{R}^n$ are the eigenvectors of Γ, and the diagonal matrix

$$\Lambda = \operatorname{diag}(\lambda_1, \lambda_2, \ldots, \lambda_n), \quad \lambda_1 \geq \lambda_2 \geq \cdots \geq \lambda_n > 0,$$

contains the eigenvalues in decreasing order. After introducing the symmetric factorization

$$\Gamma^{-1} = W\Lambda^{-1/2}\underbrace{\Lambda^{-1/2}W^{\mathsf{T}}}_{=R} = R^{\mathsf{T}}R,$$

we can write the posterior density in the form

$$\pi_{X|B}(x \mid b) \propto \exp\left(-\frac{1}{2\sigma^2}\left(\|b - Ax\|^2 + \sigma^2\|Rx\|^2\right)\right),$$

from which it can be easily seen that the maximum a posteriori (MAP) estimate is the Tikhonov regularized solution with $\alpha = \sigma^2$, and

$$R = \Lambda^{-1/2}W^{\mathsf{T}}. \tag{9.9}$$

This formulation lends itself naturally to a geometric interpretation of the Tikhonov penalty. Since

$$Rx = \sum_{j=1}^{n} \frac{w_j^{\mathsf{T}}x}{\sqrt{\lambda_j}}e_j,$$

where e_j is the jth canonical basis vector of $\mathbb{R}^n$, by penalizing the solution for the growth of those components $w_j^{\mathsf{T}}x$ that are divided by small numbers $\sqrt{\lambda_j}$, the penalty term effectively pushes the solution toward the eigenspace corresponding to *largest* eigenvalues of the covariance matrix.

Example 9.2 We illustrate the iterative solver approach by considering a small example with a forward map $A \in \mathbb{R}^{2\times 2}$. The singular value decomposition of the forward map A is assumed to be given by

$$A = UDV^{\mathsf{T}}, \tag{9.10}$$

where

$$U = \begin{bmatrix} u_1 & u_2 \end{bmatrix} = \begin{bmatrix} 1 & 0 \\ 0 & 1 \end{bmatrix}, \quad V = \begin{bmatrix} u_1 & u_2 \end{bmatrix} = \begin{bmatrix} \cos(\pi/6) & -\sin(\pi/6) \\ \sin(\pi/6) & \cos(\pi/6) \end{bmatrix},$$

and

$$D = \begin{bmatrix} \lambda_1 & \\ & \lambda_2 \end{bmatrix} = \begin{bmatrix} 0.5 & \\ & 0.1 \end{bmatrix}.$$

We choose a point $x^* \in \mathbb{R}^2$ and let b^* denote the noiseless data

$$b^* = Ax^*, \quad x^* = \begin{bmatrix} 1 \\ 1.5 \end{bmatrix}.$$

To visualize the mapping properties of A, in the top row of Fig. 9.5, we show a unit disc around x^* and the singular vectors v_j, and the image of the disc under the matrix A, which is an ellipse around b^* with semiaxes parallel to u_j. Conversely, if we consider a disc around b^*, its preimage is an ellipse around x^* with semiaxes parallel to singular vectors v_j, but this time the longer semiaxis corresponds to the smaller singular value. To better understand the effect of the different singular values, we pick two points b_1 and b_2 which can be interpreted as two noisy realizations of b with the same noise level, and map them to the x-space. We observe that, while the noise level is the same, the effect of the noise on the retrieved x-values is different, as the mapping A is more sensitive to noise in the u_2-direction than in the u_1-direction. Therefore, ideally, an inversion algorithm should treat the singular directions differently. It turns out that CGLS does exactly that, as the following simulation shows.

We start by generating a cloud of $N = 50$ perturbed data vectors around the generative model b^*,

$$b^{(j)} = b^* + r\, w^{(j)}, \quad w^{(j)} \sim \mathcal{N}(0, I_2), \quad r = 1/2.$$

In the first simulation, we perform two steps of CGLS without preconditioning, thus generating vectors

$$\{x_1^{(1)}, x_1^{(2)}, \ldots, x_1^{(N)}\} \quad \text{(one CGLS iteration)},$$
$$\{x_2^{(1)}, x_2^{(2)}, \ldots, x_2^{(N)}\} \quad \text{(two CGLS iterations)}.$$

Observe that since A is invertible, CGLS converges in two steps and we have $x_2^{(j)} = A^{-1}b^{(j)}$. In Fig. 9.6, top row, we have plotted the original data points (left panel), and the clouds of the first and second iterations (middle and right panels, respectively).

To understand the result, recall that small perturbations of x in the direction v_1 have the largest effect on the data, as the top row of Fig. 9.5 indicates. Therefore, in order to minimize the discrepancy in the first iteration step, the CGLS algorithm selects a search direction in which the forward map is most sensitive to perturbations

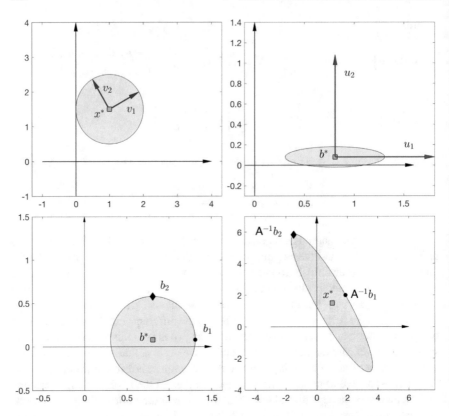

Fig. 9.5 Top row: Unit disc around the point x^* with the right singular vectors (left), and the image of it under the mapping A (right), with the left singular vectors. Bottom row: A disc of radius $r = 0.5$ around the point b^* with two points obtained by moving into the directions of the singular vectors (left), and the preimage of it by the mapping A (right)

of x. Such direction is near the singular vector v_1, and hence it is not a surprise that the first iteration cloud lies near the line from the origin (initial point of the iteration) in the direction of the first singular vector as shown in the top panel in the middle. The second iteration step resolves the rest of the discrepancy, and we arrive at the figure in the top right panel.

Next, we add a prior. Let the covariance matrix of the Gaussian prior be given in terms of its eigenvalue decomposition,

$$\Gamma = W\Lambda W^\mathsf{T},$$

where

$$W = \begin{bmatrix} w_1 & w_2 \end{bmatrix} = \begin{bmatrix} \cos\theta & -\sin\theta \\ \sin\theta & \cos\theta \end{bmatrix},$$

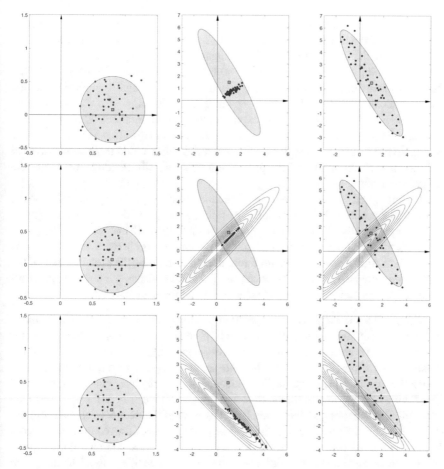

Fig. 9.6 Three runs of CGLS starting with the same data (left column). Panels in the middle column show the results after one iterations, and those in the right column after two iterations. In the top row, no preconditioner was included, while in the middle and in the bottom row, the preconditioner corresponds to the prior density plotted in the figure

and Λ is a diagonal matrix of prior variances,

$$\Lambda = \begin{bmatrix} \lambda_1^2 & \\ & \lambda_2^2 \end{bmatrix}.$$

We choose first the values as

$$\theta = \frac{\pi}{4}, \quad \lambda_1 = 1.5, \quad \lambda_2 = 0.2,$$

meaning that we expect the variable X to have relatively large variance in the direction w_1 pointing from the origin into the positive quadrant, and relatively small variance in the perpendicular direction w_2. We run again the CGLS algorithm, this time with preconditioning. The center row of Fig. 9.6 shows the result. The data on the left are the same as in the previous example, and in the center and right panels, the first and the second iteration clouds are shown. The panels show also the contour plot of the prior probability density. As expected, the second iterations with and without preconditioning coincide, as the matrix A is invertible. However, the first iterations differ: In the presence of the preconditioner, the iteration cloud lies almost perfectly on the main principal axis of the prior equiprobability ellipse corresponding to the larger eigenvalue.

To further elaborate on the role of preconditioning, we swap the roles of the prior variances, which is tantamount to rotating the prior density by an angle of π, or, equivalently, setting $\theta = -\pi/4$ in the definition of the prior. Running the preconditioned CGLS with this choice leads to the results shown in the bottom row of Fig. 9.6. Again, after the first iteration, the points lie near the principal axis of the prior equiprobability ellipse, however, the likelihood pushes the points slightly off so that a compromise between the prior belief and the data is reached.

The previous example raises a general question concerning the compatibility of prior and likelihood. In the second choice of preconditioning, in particular, the prior seems to be in conflict with what the data are suggesting, and one could come to the conclusion that the second prior is in some sense wrong. However, it is important to remember what the prior and the likelihood mean. We may have good reasons, based on previous observations or theoretical reasoning to set the prior as it was given in the example. Furthermore, in the example, we generated the data corresponding to the forward model that was used in the inverse problem. Such procedure is sometimes referred to as an *inverse crime*. In real-world applications, the data do not come from a model and can be in conflict with the model we decide to use. The mismatch between the model and data is called *model discrepancy*, and it is not a trivial matter to try to take it into account in computations.

The examples above serve as an intuitive basis for the more systematic development of the methodology.

9.4.2   Priorconditioners: Specially Chosen Preconditioners

After the preliminary considerations with low-dimensional problems of the previous section, we now develop further the idea of preconditioning with Krylov subspace solvers based on Bayesian approach. We start by summarizing the basic idea as follows.

(a) Given a linear forward model

$$B = AX + E, \quad A \in \mathbb{R}^{m \times n},$$

and assuming that

$$X \sim \mathcal{N}(0, \Gamma), \quad E \sim \mathcal{N}(0, \Sigma),$$

where $\Gamma \in \mathbb{R}^{n \times n}$ and $\Sigma \in \mathbb{R}^{m \times m}$ are symmetric positive define matrices, the goal is to approximate the MAP estimate.

(b) Using the symmetric factorization of the precision matrices,

$$\Gamma^{-1} = \mathsf{L}^{\mathsf{T}}\mathsf{L}, \quad \Sigma^{-1} = \mathsf{S}^{\mathsf{T}}\mathsf{S}, \tag{9.11}$$

write the objective functional as

$$
\begin{aligned}
\mathcal{E}(x, b) &= \frac{1}{2}(b - \mathsf{A}x)^{\mathsf{T}}\Sigma^{-1}(b - \mathsf{A}x) + \frac{1}{2}x^{\mathsf{T}}\Gamma^{-1}x \\
&= \frac{1}{2}\left(\|\mathsf{S}(b - \mathsf{A}x)\|^2 + \|\mathsf{L}x\|^2\right) \\
&= \frac{1}{2}\left(\|y - \mathsf{S}\mathsf{A}\mathsf{L}^{-1}w\|^2 + \|w\|^2\right), \quad w = \mathsf{L}x, \quad y = \mathsf{S}b.
\end{aligned}
$$

The minimization of this expression with respect to w is a standard form Tikhonov regularized problem.

(c) Replace the Tikhonov regularized problem with an approximate CGLS solution to the linear problem

$$y = \mathsf{S}\mathsf{A}\mathsf{L}^{-1}w, \quad w = \mathsf{L}x,$$

regularizing the problem with an early stopping rule.

Since the preconditioners have been constructed on the basis of a Bayesian model, the matrices S and L are referred to as the left and the right *priorconditioners*, respectively.

The approach outlined above produces a computationally efficient approximation of the MAP estimate, and we refer to this solution as the *quasi-MAP (qMAP) estimate*.

The functionality of the above approach will be analyzed more in detail below and illustrated via computed examples, but before that, the stopping rule needs to be set.

Let us start considering plain CGLS without priorconditioners to find an approximate solution to the problem

$$b = \mathsf{A}x + e,$$

where we assume that the noise is whitened, that is, e is a realization of Gaussian white noise $E \sim \mathcal{N}(0, \mathsf{I}_m)$. The CGLS iterate x_j approximating the least squares solutions satisfies

$$x_j = \operatorname{argmin}\{\|b - \mathsf{A}x\| \mid x \in \mathcal{K}_j(\mathsf{A}^{\mathsf{T}}b, \mathsf{A}^{\mathsf{T}}\mathsf{A})\},$$

where the jth Krylov subspace associated with the method is

$$\mathcal{K}_j(\mathsf{A}^\mathsf{T} b, \mathsf{A}^\mathsf{T} \mathsf{A}) = \mathrm{span}\{\mathsf{A}^\mathsf{T} b, (\mathsf{A}^\mathsf{T} \mathsf{A})\mathsf{A}^\mathsf{T} b, \ldots, (\mathsf{A}^\mathsf{T} \mathsf{A})^{j-1}\mathsf{A}^\mathsf{T} b\}.$$

Since $\mathcal{K}_j(\mathsf{A}^\mathsf{T} b, \mathsf{A}^\mathsf{T} \mathsf{A}) \subset \mathcal{K}_{j+1}(\mathsf{A}^\mathsf{T} b, \mathsf{A}^\mathsf{T} \mathsf{A})$ for all j, the norms of the discrepancies $r_j = b - \mathsf{A} x_j$ form a non-increasing sequence,

$$\|b - \mathsf{A} x_j\| \geq \|b - \mathsf{A} x_{j+1}\| \text{ for all } j.$$

To find a feasible stopping criterion similar to Morozov's discrepancy principle, observe that from the assumption about the noise,

$$\mathrm{E}\big(\|E\|^2\big) = \mathrm{E}\big(E^\mathsf{T} E\big) = \mathrm{trace}\big(\mathrm{E}\{E E^\mathsf{T}\}\big) = \mathrm{trace}(\mathsf{I}_m) = m,$$

that is, the expected size of the squared norm of the error in the data is of the order m. This suggests that the iterations should be stopped as soon as the approximate solution x_j satisfies

$$\|b - \mathsf{A} x_j\|^2 < m.$$

Typically, the stopping condition is satisfied in a number of iterations much smaller than m, making the computations much less costly than, e.g., using Tikhonov regularization.

To understand the limitations of the iterative regularization, consider the orthogonal decomposition based on the four fundamental subspaces,

$$\mathbb{R}^n = \mathcal{N}(\mathsf{A}) \oplus \mathcal{R}(\mathsf{A}^\mathsf{T}).$$

By the definition of the Krylov subspspaces,

$$\mathcal{K}_j(\mathsf{A}^\mathsf{T} b, \mathsf{A}^\mathsf{T} \mathsf{A}) \subset \mathcal{R}(\mathsf{A}^\mathsf{T}),$$

which implies that

$$x_j \perp \mathcal{N}(\mathsf{A}) \quad \text{for all } j,$$

that is, the iterates x_j never contain any information about the part of the solution that is in the null space of A. This feature may have a dramatic effect on the quality of the approximate solutions, as will be demonstrated by computed examples.

Consider now the case in which the noise and the prior covariances are not necessarily identity matrices. Our main interest here is on the effect of the prior, therefore, we assume that $\Sigma = \mathsf{I}_m$. If this is not the case, we can always reduce the problem to this case by whitening, that is, by replacing

$$b \to \mathsf{S} b, \quad \mathsf{A} \to \mathsf{S} \mathsf{A},$$

where $\mathsf{S} \in \mathbb{R}^{m \times m}$ is a symmetric factor of the noise precision matrix Σ^{-1}. Introducing the notation $\widetilde{\mathsf{A}} = \mathsf{A} \mathsf{L}^{-1}$, we define the qMAP estimate x_j by

$$w_j = \mathrm{argmin}\{\|b - \widetilde{\mathbf{A}}w\| \mid x \in \mathcal{K}_j(\widetilde{\mathbf{A}}^\mathsf{T}b, \widetilde{\mathbf{A}}^\mathsf{T}\widetilde{\mathbf{A}})\}, \quad w_j = \mathbf{L}x_j,$$

and we stop the iterations as soon as j satisfies the stopping condition

$$\|b - \widetilde{\mathbf{A}}w_j\|^2 < m.$$

By definition of the Krylov subspaces, it follows that

$$x_j \in \mathbf{L}^{-1}\mathcal{K}_j(\widetilde{\mathbf{A}}^\mathsf{T}b, \widetilde{\mathbf{A}}^\mathsf{T}\widetilde{\mathbf{A}})$$
$$= \mathrm{span}\{\mathbf{L}^{-1}(\widetilde{\mathbf{A}}^\mathsf{T}\widetilde{\mathbf{A}})^\ell \widetilde{\mathbf{A}}b \mid 0 \le \ell \le j - 1\}.$$

Observe that, from (9.11),

$$\mathbf{L}^{-1}\widetilde{\mathbf{A}}^\mathsf{T} = \mathbf{L}^{-1}\mathbf{L}^{-\mathsf{T}}\mathbf{A}^\mathsf{T}$$
$$= (\mathbf{L}^\mathsf{T}\mathbf{L})^{-1}\mathbf{A}^\mathsf{T}$$
$$= \Gamma\mathbf{A}^\mathsf{T},$$

and inductively,

$$\mathbf{L}^{-1}(\widetilde{\mathbf{A}}^\mathsf{T}\widetilde{\mathbf{A}})^\ell \widetilde{\mathbf{A}}^\mathsf{T} = \mathbf{L}^{-1}\widetilde{\mathbf{A}}^\mathsf{T}\widetilde{\mathbf{A}}(\widetilde{\mathbf{A}}^\mathsf{T}\widetilde{\mathbf{A}})^{\ell-1}\widetilde{\mathbf{A}}^\mathsf{T}$$
$$= \Gamma\mathbf{A}^\mathsf{T}\mathbf{A}\mathbf{L}^{-1}(\widetilde{\mathbf{A}}^\mathsf{T}\widetilde{\mathbf{A}})^{\ell-1}\widetilde{\mathbf{A}}^\mathsf{T}$$
$$= (\Gamma\mathbf{A}^\mathsf{T}\mathbf{A})^\ell \Gamma\mathbf{A}^\mathsf{T}.$$

Therefore, priorconditioning by $\mathbf{L}$ modifies the subspace of the approximate solutions x_j, so that

$$x_j \in \mathrm{span}\{(\Gamma\mathbf{A}^\mathsf{T}\mathbf{A})^\ell \Gamma\mathbf{A}^\mathsf{T}b \mid 0 \le \ell \le j - 1\}.$$

In particular, we observe that

$$x_j \in \Gamma(\mathcal{R}(\mathbf{A}^\mathsf{T})) = \Gamma(\mathcal{N}(\mathbf{A})^\perp),$$

and the qMAP solution is not necessarily orthogonal to the null space, but may have a significant component in that space. In particular, by writing $\mathbf{A}$ in terms of its row vectors as

$$\mathbf{A} = \begin{bmatrix} a_{(1)}^\mathsf{T} \\ \vdots \\ a_{(m)}^\mathsf{T} \end{bmatrix}, \quad a_{(j)} \in \mathbb{R}^n,$$

we have

$$\mathcal{R}(\mathbf{A}^\mathsf{T}) = \mathrm{span}\{a_{(1)}, \ldots, a_{(m)}\},$$

while

$$\Gamma\big(\mathcal{R}(\mathsf{A}^{\mathsf{T}})\big) = \mathrm{span}\big\{\Gamma a_{(1)}, \ldots, \Gamma a_{(m)}\big\}.$$

Visualizing the basis vectors of the subspaces of the priorconditioned CGLS iterates is often an efficient way to understand the effect of priorconditioning.

The role of the null spaces is clearly highlighted in the next example. Before the example, however, we point out an important detail of the priorconditioning that is not immediately obvious. Assume that the prior covariance matrix, or its inverse is known only up to a multiplicative constant. This could be the case, e.g., when smoothness priors are used, but no other information than the order of the smoothness is available. When scaling the matrix L by a constant, $\mathsf{L} \to \mu\mathsf{L}$, the priorconditioned CGLS algorithm computes a solution that is scaled by that same constant, $w_j \to \mu w_j$. However, when we return to the original variable x_j, we write

$$x_j = \big(\mu\mathsf{L}\big)^{-1}(\mu w_j) = \mathsf{L}^{-1}w_j,$$

that is, the effect of the scaling vanishes! Therefore, the prior covariance, or its inverse, needs to be specified only up to a multiplicative constant, making the approach easier to apply in practice.

Example 9.3 In this example, we consider an underdetermined one-dimensional deconvolution problem with a Gaussian convolution kernel. We divide the unit interval in $n-1$ intervals by n equidistant points, $s_\ell = \ell/(n-1)$, $0 \le \ell \le n-1$, and approximate the convolution with the kernel a by a finite sum,

$$g(t) = \int_0^1 a(t-s)f(s)ds \approx \frac{1}{n-1}\sum_{\ell=0}^{n-1} a(t-s_\ell)f(s_\ell),$$

where the convolution kernel is

$$a(t) = e^{-t^2/(2\gamma^2)}, \quad \gamma = 0.02.$$

The data are assumed to consist of m noisy measurements at points t_j, $1 \le j \le m$,

$$b_j = g(t_j) + e_j, \quad 1 \le j \le m,$$

or, by denoting the unknown grid values of the function f by $x_j = f(s_{j-1})$, $1 \le j \le n$,

$$b = \mathsf{A}x + e, \quad \mathsf{A} \in \mathbb{R}^{m\times n}, \quad a_{jk} = \frac{1}{n-1}a(t_j - s_{k-1}).$$

We assume that the noise vector e is a realization of Gaussian scaled white noise E, $E \sim \mathcal{N}(0, \sigma^2\mathsf{I}_m)$, and we set $\sigma = 10^{-3}$. In our computed example, we set $n = 150$, and $m = 6$, choosing the observation points as $t_j = j/(m+1)$, $1 \le j \le m$. This way, the problem is strongly underdetermined, and the matrix A has a null space of dimension $n - m = 144$. We generate the data by using a smooth function,

$$f(s) = \frac{1}{2}\left(1 - \cos\frac{3\pi}{2}s\right).\qquad(9.12)$$

We demonstrate the effect of priorconditioning by choosing the prior precision matrix to correspond to a second-order smoothness with free end points,

$$\Gamma^{-1} = \mu \begin{bmatrix} \alpha & & & & \\ -1 & 2 & -1 & & \\ & \ddots & \ddots & \ddots & \\ & & -1 & 2 & -1 \\ & & & & \alpha \end{bmatrix} \in \mathbb{R}^{n\times n},$$

where $\mu > 0$ is an arbitrary scaling, and α is selected so that the marginal variances of the endpoints equal the marginal variance of the values x_j near $s = 1/2$.

We first run the plain CGLS algorithm with no priorconditioning, stopping the iterations at the discrepancy. It turns out that the norm of the discrepancy $r_j = Ax_j - b$ with whitened noise reaches the level $\sqrt{m}$ in three iterations. The iterates x_j are plotted in Fig. 9.7 together with the generative model (9.12).

The estimated solution does not correspond well to the generative model mainly because of the large null space of the forward model. Indeed, the range of the matrix A^T consists of six vectors $v^{(j)}$, the rows of A, given by

$$v_k^{(j)} = \frac{1}{n-1}e^{-(t_j - s_{k-1})^2/(2\gamma^2)}, \quad 1 \le j \le 6, \quad 1 \le k \le n,\qquad(9.13)$$

which are the narrow Gaussians plotted in Fig. 9.8, and the plain CGLS reconstruction is a linear combination of these vectors, which are not capable of representing a slowly varying smooth function like the generative model.

Fig. 9.7 The three first iterations of the CGLS algorithm with no priorconditioning. The third iteration corresponds to the solution that satisfies the discrepancy criterion

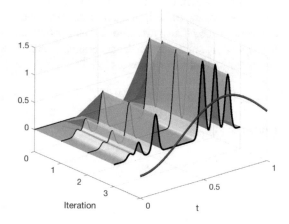

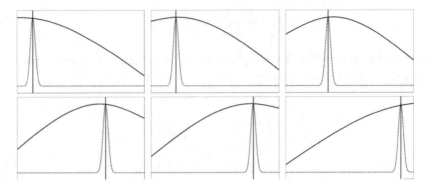

Fig. 9.8 The basis vectors $v^{(j)}$ given by (9.13), dotted line, and the basis vectors $\widetilde{v}^{(j)}$ corresponding to the priorconditioned problem (9.14), solid line

Fig. 9.9 The six first iterations of the CGLS algorithm with the priorconditioned forward map, the last iteration being the proposed solution

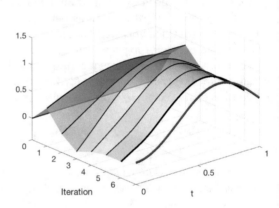

We then switch to the priorconditioned CGLS. Figure 9.9 shows the iterates, the last one satisfying the stopping criterion. We see that in order to satisfy the stopping criterion, twice as many iterations than for the plain CGLS were needed, however, the solution is quite close to the generative model. The solution in this case is a linear combination of the basis vectors

$$\widetilde{v}^{(j)} = \Gamma v^{(j)}, \quad 1 \leq j \leq m, \tag{9.14}$$

plotted in Fig. 9.8.

9.4.3 Stopping Rule Revisited

We finish this section by revisiting the stopping criterion in the presence of prior-conditioners. Recall that in the presence of a priorconditioner L, the least squares problem

$$y = SAL^{-1}w, \quad w = Lx, \quad y = Sb,$$

solved approximately using CGLS with early stopping is a proxy for the original problem of minimizing the objective function

$$\mathcal{E}(x, b) = \|SAx - y\|^2 + \|Lx\|^2.$$

When we generate the sequence of iterates

$$x_j = L^{-1}w_j, \quad w_j \in \mathcal{K}_j(\widetilde{A}^{\mathsf{T}}b, \widetilde{A}^{\mathsf{T}}\widetilde{A}),$$

it may happen that the objective function $\mathcal{E}(x_j, b)$ stops decreasing before the stopping criterion

$$\|Sb - SAx_j\|^2 < m \tag{9.15}$$

is satisfied. In other words, the stopping criterion following the Morozov discrepancy principle may be optimal for the proxy problem, while an earlier stopping would be consistent with the original problem. This consideration justifies a slight modification of the stopping criterion, whereas the CGLS iteration is stopped as soon as either the criterion (9.15) or

$$\mathcal{E}(x_{j+1}, b) > \mathcal{E}(x_j, b)$$

is met. This way, the balance between the prior and the likelihood is better respected.

Notes and Comments

The Conjugate Gradient algorithm was first introduced in the paper of Hestenes and Stiefel from 1952 [44], where the reason for the name of the algorithm is clearly explained. We refer to the book [69] for a more modern and comprehensive discussion of Krylov subspace methods.

The idea of using right preconditioners that guide the CGLS iterations to favor at first solutions with desirable features has been discussed in the deterministic context in, e.g., [42]. The connection between right preconditioners and Bayesian methods, and the term priorconditioner was first introduced in [8, 17].

The properties of the priorconditioned CGLS iterations can be analyzed by using Generalized Singular Value Decompositions (GSVD) introduced in [81]. We refer to the review article [13] for further details.

Chapter 10
Hierarchical Models and Bayesian Sparsity

Algorithms for finding sparse solutions are of importance in several application areas, and the design of penalty functionals that promote sparsity continues to be the topic of active research. In the Bayesian framework, searching for a sparse solution implicitly states that there is an a priori belief that only a few components of the unknown are nonzero. Therefore sparsity is a trait that should be described by the prior, although it is not immediately clear how that can be done in a way that leads to a computationally convenient posterior. Sparsity is more a qualitative than a quantitative trait, and in general there are no set guidelines for deciding when a solution is sparse. The concept of sparsity is dimension-dependent: In two dimensions, for example, a vector with one vanishing component may be regarded as sparse, while 50% vanishing components in an unknown with a million of components may not qualify it as sparse. Moreover, the concept is complicated by the fact that sparsity is not a property of the unknown alone, but of its representation: *Sparse coding* is a process of representing the unknown in a suitable basis or frame in terms of few nonzero coefficients. To elucidate the latter statement, a discretized sinusoidal arc is not a sparse vector, however, its spectrum consists of just one component and in the spectral basis it is indeed sparse.

In this chapter, we propose a family of conditionally Gaussian hierarchical priors that are particularly well suited to express sparsity belief, and that in combination with a Gaussian likelihood yield posteriors whose maximizers can be computed efficiently even for high-dimensional problems using Krylov subspace linear solvers.

10.1 Posterior Densities with Conditionally Gaussian Priors

Consider linear inverse problems with additive Gaussian noise,

$$b = Ax + e, \quad A \in \mathbb{R}^{m \times n},$$

© The Author(s), under exclusive license to Springer Nature Switzerland AG 2023
D. Calvetti and E. Somersalo, *Bayesian Scientific Computing*, Applied Mathematical
Sciences 215, https://doi.org/10.1007/978-3-031-23824-6_10

where the noise $e \in \mathbb{R}^m$ is a realization of a Gaussian random variable $E \sim \mathcal{N}(0, \Sigma)$, with a symmetric positive definite matrix $\Sigma \in \mathbb{R}^{m \times m}$. If

$$\Sigma^{-1} = \mathsf{S}^\mathsf{T}\mathsf{S}$$

is a symmetric factorization of the precision matrix of the noise, a multiplication of the stochastic extension of the problem from the left by S,

$$\mathsf{S}B = \mathsf{S}\mathsf{A}X + \mathsf{S}E,$$

whitens the noise, since $\mathsf{S}E \sim \mathcal{N}(0, \mathsf{I}_m)$. Therefore, without of loss of generality, we restrict our attention to the whitened problem

$$B = \mathsf{A}X + E, \quad E \sim \mathcal{N}(0, \mathsf{I}_m).$$

We want to design a prior that favors a solution X with most of its entries zero or close to zero. We recall the discussion about the hierarchical prior models from Chap. 6, and develop further the ideas here.

As in the earlier discussion, we consider an uncorrelated conditionally Gaussian prior model for X,

$$
\pi_{X|\Theta}(x \mid \theta) = \prod_{j=1}^{n} \frac{1}{\sqrt{2\pi\theta_j}} \exp\left(-\frac{1}{2}\frac{x_j^2}{\theta_j}\right)
$$
$$
= \left(\frac{1}{2\pi}\right)^{n/2} \exp\left(-\frac{1}{2}\sum_{j=1}^{n}\frac{x_j^2}{\theta_j} - \frac{1}{2}\sum_{j=1}^{n}\log\theta_j\right),
$$

that depends on the vector θ of the variances of the components. If the vector θ is not known, we model it as a random variable Θ with independent components, and we express our a priori belief about Θ_j in terms of the gamma distribution,

$$
\Theta_j \sim \text{Gamma}(\beta, \vartheta_j), \quad \text{or} \quad \pi_{\Theta_j}(\theta_j) = \frac{\theta_j^{\beta-1}}{\Gamma(\beta)\vartheta_j^{\beta}} e^{-\theta_j/\vartheta_j},
$$

where $\beta > 0$ is the shape parameter, and ϑ_j is a scale parameter. We assume that $\beta > 0$ is the same for every component, while we allow the values of the scale parameters to be different. This flexibility with the values of the scale parameters will turn out to be quite convenient, as we will see later. Assuming that the variances Θ_j are mutually independent, we arrive at the hyperprior model

$$
\pi_{\Theta}(\theta) = \frac{1}{\Gamma(\beta)^n} \frac{1}{(\vartheta_1 \cdots \vartheta_n)^\beta} \exp\left(-\sum_{j=1}^{n}\frac{\theta_j}{\vartheta_j} + (\beta-1)\sum_{j=1}^{n}\log\theta_j\right).
$$

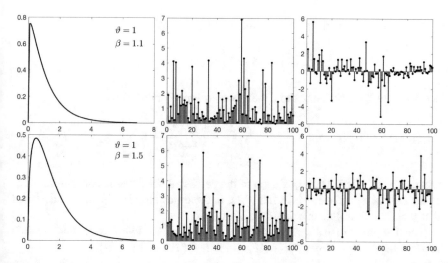

Fig. 10.1 Gamma distributions (left) with two different choices of the shape parameter. In the middle, realizations of $\Theta \in \mathbb{R}^{100}$ with independent components distributed according to the gamma distribution, and on the right, corresponding realizations of $X \in \mathbb{R}^{100}$ with independent components $X_j \sim \mathcal{N}(0, \theta_j)$

The joint prior model for the pair (X, Θ) can now be written as

$$\pi_{X,\Theta}(x, \theta) = \pi_{X|\Theta}(x \mid \theta)\pi_\Theta(\theta)$$

$$= C(\beta, \vartheta)\exp\left(-\frac{1}{2}\sum_{j=1}^{n}\frac{x_j^2}{\theta_j} + (\beta - \frac{3}{2})\sum_{j=1}^{n}\log\theta_j - \sum_{j=1}^{n}\frac{\theta_j}{\vartheta_j}\right),$$

where the normalizing constant is given by

$$C(\beta, \vartheta) = \frac{1}{(\sqrt{2\pi}\,\Gamma(\beta))^n}\frac{1}{(\vartheta_1\cdots\vartheta_n)^\beta}. \tag{10.1}$$

Before considering the posterior distribution, we demonstrate the properties of the hierarchical prior model by showing random draws from it. In Fig. 10.1, the panels in the left column show the gamma distributions with two different shape parameter values, $\beta = 1.1$ and $\beta = 1.5$. In the middle column, we show realizations θ of the variance vector $\Theta \in \mathbb{R}^{100}$ with independent components following the gamma distribution, and finally, on the right, draws of the vector $X \in \mathbb{R}^{100}$ with independent components $X_j \sim \mathcal{N}(0, \theta_j)$ are shown. The draws of Θ indicate that the relatively fat tail of the gamma distribution allows occasional outliers of the variances, leading to realizations of X in which small values are favored, while occasional relatively large values are possible.

Now, we combine the prior with the likelihood density,

$$\pi_{B|X}(b \mid x) \propto \exp\left(-\frac{1}{2}\|b - Ax\|^2\right),$$

according to Bayes' formula to get the posterior probability density

$$\pi_{X,\Theta|B}(x, \theta \mid b) \propto \exp\left(-\frac{1}{2}\|b - Ax\|^2 - \frac{1}{2}\sum_{j=1}^n \frac{x_j^2}{\theta_j} + \eta \sum_{j=1}^n \log \theta_j - \sum_{j=1}^n \frac{\theta_j}{\vartheta_j}\right),$$

where $\eta = \beta - 3/2$. The expression of the neglected normalizing constant in terms of the parameters β and ϑ is as in (10.1). In the following discussion, it is assumed that $\beta > 3/2$, thus guaranteeing that $\eta > 0$.

The first problem that we address is the following. *Given the observation $b \in \mathbb{R}^m$, find the maximum a posteriori (MAP) estimate for the pair (X, Θ),*

$$(x, \theta)_{\text{MAP}} = \text{argmax} \, \pi_{X,\theta|B}(x, \theta \mid b).$$

In the search for a maximizer, we take advantage of the fact that a maximizer minimizes the negative logarithm of the posterior density, or *Gibbs energy* functional,

$$\mathscr{E}(x, \theta) = \overbrace{\frac{1}{2}\|b - Ax\|^2 + \underbrace{\frac{1}{2}\sum_{j=1}^n \frac{x_j^2}{\theta_j}}_{} - \eta \sum_{j=1}^n \log \theta_j + \sum_{j=1}^n \frac{\theta_j}{\vartheta_j}}^{(a)}. \qquad (10.2)$$

$$\underbrace{\phantom{\frac{1}{2}\sum \frac{x_j^2}{\theta_j} - \eta \sum \log \theta_j + \sum \frac{\theta_j}{\vartheta_j}}}_{(b)}$$

The structure of the Gibbs energy functional suggests a minimization strategy based on the following sequential alternating scheme:

(a) Fix θ, and minimize the x-dependent part (a) of $\mathscr{E}(x, \theta)$ with respect to x;
(b) Fix x to the updated value, and minimize the θ-dependent part (b) of $\mathscr{E}(x, \theta)$ with respect to θ.

To implement this algorithm, consider the two minimization problems separately. Introducing a notation similar to that for conditional probability, we write the expressions (a) and (b) as

$$\mathscr{E}_\theta(x) = \frac{1}{2}\|b - Ax\|^2 + \frac{1}{2}\sum_{j=1}^n \frac{x_j^2}{\theta_j},$$

$$\mathscr{E}_x(\theta) = \frac{1}{2}\sum_{j=1}^n \frac{x_j^2}{\theta_j} - \eta \sum_{j=1}^n \log \theta_j + \sum_{j=1}^n \frac{\theta_j}{\vartheta_j}.$$

To minimize $\mathscr{E}_\theta(x)$ with θ fixed, observe that this is a quadratic expression in x and can be written as

$$\mathscr{E}_\theta(x) = \frac{1}{2}\|b - Ax\|^2 + \frac{1}{2}\|D_\theta^{-1/2}x\|^2 = \frac{1}{2}\left\|\begin{bmatrix} A \\ D_\theta^{-1/2} \end{bmatrix} x - \begin{bmatrix} b \\ 0 \end{bmatrix}\right\|^2, \quad (10.3)$$

where D_θ is the diagonal matrix

$$D_\theta = \begin{bmatrix} \theta_1 & & \\ & \ddots & \\ & & \theta_n \end{bmatrix} \in \mathbb{R}^{n \times n}.$$

Therefore, solving the minimization problem amounts to solving a linear least squares problem.

The expression $\mathscr{E}_x(\theta)$, on the other hand, is a sum of terms each depending only on a single component θ_j, therefore the minimization can be carried out independently for each component. Writing the first-order optimality condition with respect to each θ_j,

$$\frac{\partial}{\partial \theta_j}\mathscr{E}_x(\theta) = -\frac{1}{2}\frac{x_j^2}{\theta_j^2} - \eta\frac{1}{\theta_j} + \frac{1}{\vartheta_j} = 0,$$

and multiplying it by θ_j^2/ϑ_j yields the second-order equation

$$\left(\frac{\theta_j}{\vartheta_j}\right)^2 - \eta\frac{\theta_j}{\vartheta_j} - \frac{1}{2}\frac{x_j^2}{\vartheta_j} = 0,$$

with the unique positive solution

$$\theta_j = \vartheta_j\left(\frac{\eta}{2} + \sqrt{\frac{\eta^2}{4} + \frac{x_j^2}{2\vartheta_j}}\right) = f(x_j, \vartheta_j, \eta). \quad (10.4)$$

Therefore the update of θ can be done in a component-wise manner by a closed form formula evaluation.

To put the above idea on more solid footing, we prove that the function $\mathscr{E}(x, \theta)$ has a global unique minimizer by proving that the function is strictly convex. This is done by showing that the Hessian matrix is symmetric positive definite. We begin by partitioning the Hessian of $\mathscr{E}(x, \theta)$ as

$$H = \begin{bmatrix} \nabla_x\nabla_x\mathscr{E}(x, \theta) & \nabla_x\nabla_\theta\mathscr{E}(x, \theta) \\ \nabla_\theta\nabla_x\mathscr{E}(x, \theta) & \nabla_\theta\nabla_\theta\mathscr{E}(x, \theta) \end{bmatrix} \in \mathbb{R}^{2n \times 2n},$$

and compute the four blocks separately.

It follows from (10.3) that

$$\nabla_x\nabla_x\mathscr{E}(x, \theta) = A^\mathsf{T}A + D_\theta^{-1}.$$

Likewise, a simple computation shows that

$$\nabla_\theta \nabla_x \mathcal{E}(x, \theta) = - \begin{bmatrix} x_1/\theta_1^2 & & \\ & \ddots & \\ & & x_n/\theta_n^2 \end{bmatrix},$$

while

$$\nabla_\theta \nabla_\theta \mathcal{E}(x, \theta) = \begin{bmatrix} x_1^2/\theta_1^3 + \eta/\theta_1^2 & & \\ & \ddots & \\ & & x_n^2/\theta_n^3 + \eta/\theta_n^2 \end{bmatrix}.$$

To show the positive definiteness of H, let

$$q = \begin{bmatrix} u \\ v \end{bmatrix} \in \mathbb{R}^{2n},$$

and evaluate the quadratic form,

$$q^\mathsf{T} \mathsf{H} q = u^\mathsf{T}(\mathsf{A}^\mathsf{T}\mathsf{A} + \mathsf{D}_\theta^{-1})u - 2u^\mathsf{T} \begin{bmatrix} x_1/\theta_1^2 & & \\ & \ddots & \\ & & x_n/\theta_n^2 \end{bmatrix} v$$

$$+ v^\mathsf{T} \begin{bmatrix} x_1^2/\theta_1^3 + \eta/\theta_1^2 & & \\ & \ddots & \\ & & x_n^2/\theta_n^3 + \eta/\theta_n^2 \end{bmatrix} v$$

$$= \|\mathsf{A}u\|^2 + \sum_{j=1}^n \left(\frac{u_j^2}{\theta_j} - 2\frac{u_j x_j v_j}{\theta_j^2} + \frac{v_j^2 x_j^2}{\theta_j^3} + \eta\frac{v_j^2}{\theta_j} \right)$$

$$= \|\mathsf{A}u\|^2 + \sum_{j=1}^n \frac{1}{\theta_j} \left(\left(u_j - \frac{v_j x_j}{\theta_j}\right)^2 + \eta v_j^2 \right) \geq 0,$$

with equality holding only if $u = v = 0$, that is, $q = 0$. Therefore, since H is positive definite, $\mathcal{E}(x, \theta)$ is convex. Moreover, one can check that the Gibbs energy functional tends to infinity at the limits $\theta_j \to 0+$ or $\theta \to \infty$, as well as when $\|x\| \to \infty$. We therefore conclude that its minimizer exists and is unique.

It follows from the convexity of the objective function, that the following algorithm converges to the unique global minimizer of the objective function $\mathcal{E}(x, \theta)$, or, equivalently, to the unique MAP estimate of the posterior density.

Iterative Alternating Sequential (IAS) algorithm for MAP estimation: Given the shape and scale parameters $\beta > 3/2$, $\vartheta \in \mathbb{R}_+^n$ and data $b \in \mathbb{R}^m$,

1. **Initialize:** Set $\theta = \vartheta$.

2. **Iterate** Until a convergence criterion is satisfied:

   (a) Update x, solving the least squares problem

$$x = \operatorname{argmin} \left\| \begin{bmatrix} \mathsf{A} \\ \mathsf{D}_\theta^{-1/2} \end{bmatrix} x - \begin{bmatrix} b \\ 0 \end{bmatrix} \right\|^2 .$$

   (b) Update θ, setting

$$\theta_j = \vartheta_j \left(\frac{\eta}{2} + \sqrt{\frac{\eta^2}{4} + \frac{x_j^2}{2\vartheta_j}} \right), \quad 1 \le j \le n. \tag{10.5}$$

While the IAS algorithm itself is rather simple, the objective function being mini-mized depends on the values of the shape and scale parameters. Moreover, a stopping rule terminating the iterations must be specified, and in order to use it for large-scale problems, a computationally efficient solution of the least squares problem is needed. Furthermore, the claim that the hierarchical model favors sparse solutions was rather heuristic and requires a more rigorous justification. The latter question is also related to the choice of the shape and scaling parameters. We address these points in detail in the next subsection.

10.1.1   IAS, Sparsity, Sensitivity Weighting and Exchangeability

To provide a strategy for setting the values of the parameters of the hyperprior, we consider their effect on the MAP estimate. To understand the role of β, consider the objective function $\mathscr{E}(x, \theta)$ when (x, θ) is on the manifold

$$\mathcal{M}_\eta = \{(x, \theta) \mid \theta = f(x, \vartheta, \eta)\} \subset \mathbb{R}^{2n},$$

where $f : \mathbb{R}^n \to \mathbb{R}^n$ is defined component-wise by the formula (10.4). Note that each iteration round of the IAS algorithm determines an iterate that lies on this manifold. In particular, we are interested in the behavior of the objective function as $\eta \to 0+$. Therefore, consider each term in the objective function $\mathscr{E}(x, \theta)$ restricted to $\mathcal{M}_\eta$. The first term is

$$\frac{1}{2} \frac{x_j^2}{\theta_j} = \frac{1}{2} \frac{x_j^2}{f(x_j, \vartheta_j, \eta)} = \frac{1}{2} \frac{x_j^2}{\vartheta_j \left(\frac{\eta}{2} + \sqrt{\frac{\eta^2}{4} + \frac{x_j^2}{2\vartheta_j}} \right)},$$

which, in the limit as $\eta \to 0+$, converges to

$$\lim_{\eta \to 0+} \frac{1}{2} \frac{x_j^2}{f(x_j, \vartheta_j, \eta)} = \frac{1}{\sqrt{2}} \frac{|x_j|}{\sqrt{\vartheta_j}}.$$

Similarly,

$$\lim_{\eta \to 0} \frac{\theta_j}{\vartheta_j} = \lim_{\eta \to 0} \frac{f(x_j, \vartheta_j, \eta)}{\vartheta_j} = \lim_{\eta \to 0} \left(\frac{\eta}{2} + \sqrt{\frac{\eta^2}{4} + \frac{x_j^2}{2\vartheta_j}} \right) = \frac{1}{\sqrt{2}} \frac{|x_j|}{\sqrt{\vartheta_j}},$$

and

$$\lim_{\eta \to 0+} \eta \log \theta_j = \lim_{\eta \to 0+} \eta \log f(x_j, \vartheta_j, \eta) = 0.$$

Therefore, we conclude that

$$\lim_{\eta \to 0+} \mathscr{E}\left(x, f(x, \vartheta, \eta)\right) = \frac{1}{2} \|b - Ax\|^2 + \sqrt{2} \sum_{j=1}^{n} \frac{|x_j|}{\sqrt{\vartheta_j}}, \tag{10.6}$$

which is equivalent to saying that, for small $\eta > 0$, the objective function, restricted to the manifold $\mathcal{M}_\eta$, approaches the mean square error term augmented with a weighted ℓ_1-penalty term. The fact that penalizing growth in the ℓ_1-norm favors solutions with small support suggests that when η is small, the hierarchical model with gamma hyperprior favors sparse solutions. Moreover, it can be shown that the solution of the IAS algorithm converges to the minimizer of (10.6). We omit the proof here; references to the proof and further analysis of the algorithm can be found in Notes and Comments at the end of the chapter. We conclude that the value of β should be chosen to reflect the level of sparsity that we expect a priori.

To select the values of the scale parameters, consider first a reduced model in which $\vartheta_j = \vartheta_0$, that is, all components are equal. To understand what the parameter ϑ_0 represents, we consider the updating formula (10.5). If $x_j = 0$, we get the lower bound for θ_j,

$$\theta_j = \eta \vartheta_j = \eta \vartheta_0,$$

therefore the product $\eta \vartheta_0$ represents the variance of the components outside the support of x, thus if a clean background is desired, its value should be small. On the other hand, making ϑ_0 too small may have an adverse effect on the dynamic range of the solution, as θ_j represents the prior variance of x_j. These considerations will be tested with computed examples at the end of this chapter.

In the search for an automatic way of setting the value of the scale parameters ϑ, we recall that in inverse problems, the selection of the value of the Tikhonov regularization parameter depends on the noise level. Therefore it is not too surprising that the noise in the data plays a role also in the Bayesian approach. In preparation for the discussion of the role of ϑ, we start by defining what we mean by the *signal-to-noise ratio* (SNR) . For simplicity, we limit our discussion to a linear observation model of the form

$$B = AX + E.$$

The SNR is defined as

$$SNR = \frac{\mathrm{E}\big(\|B\|^2\big)}{\mathrm{E}\big(\|E\|^2\big)},$$

or the ratio of the signal power and the noise power. In engineering literature, the SNR is often given in decibels,

$$SNR_{\mathrm{dB}} = 10 \log_{10} SNR.$$

To relate the SNR to the prior and likelihood models, we observe that

$$\mathrm{E}\big(\|E\|^2\big) = \sum_{j=1}^{m} \mathrm{E}\big(E_j^2\big) = \mathrm{trace}\,\big(\mathrm{E}\big(EE^{\mathsf{T}}\big)\big) = \mathrm{trace}(\Sigma).$$

Similarly,

$$
\begin{aligned}
\mathrm{E}\big(\|B\|^2\big) &= \mathrm{trace}\,\big(\mathrm{E}\big(BB^{\mathsf{T}}\big)\big) \\
&= \mathrm{trace}\,\big(\mathrm{E}\big((AX+E)(AX+E)^{\mathsf{T}}\big)\big) \\
&= \mathrm{trace}\,\big(\mathrm{E}\big(AX(AX)^{\mathsf{T}}\big)\big) + \mathrm{trace}\,\big(\mathrm{E}\big(EE^{\mathsf{T}}\big)\big) \\
&= \mathrm{trace}\,\big(A\mathrm{E}\big(XX^{\mathsf{T}}\big)A^{\mathsf{T}}\big) + \mathrm{trace}(\Sigma).
\end{aligned}
$$

To calculate the expectation above, we write it in terms of the probability densities,

$$
\begin{aligned}
\mathrm{E}\big(XX^{\mathsf{T}}\big) &= \int \int xx^{\mathsf{T}} \pi_{X,\Theta}(x,\theta)dxd\theta \\
&= \int \left(\int xx^{\mathsf{T}} \pi_{X|\Theta}(x \mid \theta)dx \right) \pi_{\Theta}(\theta)d\theta \\
&= \int \mathsf{D}_{\theta} \pi_{\Theta}(\theta)d\theta \\
&= \beta \mathsf{D}_{\vartheta},
\end{aligned}
$$

since the expectation of $\Theta_j \sim \mathrm{Gamma}(\beta, \vartheta_j)$ is $\beta\vartheta_j$, which is easy to check using the properties of gamma functions: By a change of variables, with $t = \theta_j/\vartheta_j$,

$$
\begin{aligned}
\mathrm{E}(\Theta_j) &= \frac{1}{\vartheta_j^{\beta}} \int_0^{\infty} \theta_j \theta_j^{\beta-1} e^{-\theta_j/\vartheta_j} d\theta_j \\
&= \frac{\vartheta_j}{\Gamma(\beta)} \int_0^{\infty} t^{\beta} e^{-t} dt \\
&= \frac{\vartheta_j}{\Gamma(\beta)} \Gamma(\beta+1) \\
&= \beta \vartheta_j.
\end{aligned}
$$

By expressing the matrix A in terms of its columns,

$$
\mathsf{A} = \begin{bmatrix} a_1 \cdots a_n \end{bmatrix}, \quad a_j \in \mathbb{R}^m,
$$

we have

$$
\mathsf{A} \mathsf{D}_{\vartheta} \mathsf{A}^{\mathsf{T}} = \sum_{j=1}^{n} \vartheta_j a_j a_j^{\mathsf{T}}.
$$

Taking the trace of both sides,

$$
\mathrm{trace}(\mathsf{A} \mathsf{D}_{\vartheta} \mathsf{A}^{\mathsf{T}}) = \sum_{j=1}^{n} \vartheta_j \mathrm{trace}(a_j a_j^{\mathsf{T}}) = \sum_{j=1}^{n} \vartheta_j \|a_j\|^2,
$$

and substituting this expression in the formula for the SNR, we obtain

$$
SNR = \frac{\beta \sum_{j=1}^{n} \vartheta_j \|a_j\|^2}{\mathrm{trace}(\Sigma)} + 1.
$$

So far, sparsity has not played any role. To introduce the sparsity, suppose, for the moment, that we know the support S of X, defined as a subset of the index set $\{1, 2, \ldots, n\}$ by

$$
j \in S = \mathrm{supp}(X) \text{ if and only if } X_j \neq 0,
$$

where $X_j \neq 0$ means that X_j is not a zero random variable, although some of its realizations may assume the value zero. If S is known, we can define the SNR conditioned on the support of the signal as

$$
SNR_S = \frac{\beta \sum_{j \in S} \vartheta_j \|a_j\|^2}{\mathrm{trace}(\Sigma)} + 1. \tag{10.7}
$$

According to this definition, there is no contribution to the signal from components outside the support, which are known to vanish.

Consider now a case in which the support is not known, but we know its cardinality,

$$
\|X\|_0 = \#\mathrm{supp}(X) = \text{number of non-vanishing components of } X.
$$

We introduce now a concept of $SNR - exchangeability$. We recall that in probability theory, exchangeability of a multivariate random variable means that the probability density is invariant under permutations of the coordinates. In analogy with
that, we assume that, if the cardinality of the support is known, the SNR is invariant
with respect to permutations of the coordinates, that is

$$SNR_S = SNR_{S'} \text{ whenever } \#S = \#S'.$$

In light of the formula (10.7), this is tantamount to requiring that

$$\sum_{j \in S} \vartheta_j \|a_j\|^2 = \sum_{j \in S'} \vartheta_j \|a_j\|^2 \text{ whenever } \#S = \#S'.$$

Therefore, a way to express our belief that any group of $\#S$ nonzero entries should
have the same possibility of yielding the given SNR is to set

$$\vartheta_j = \frac{C}{\|a_j\|^2}, \quad 1 \le j \le n.$$

To determine the value of the scalar C, observe that with $\#S = k$, we have now

$$SNR_S = \frac{\beta \sum_{j \in S} \vartheta_j \|a_j\|^2}{\text{trace}(\Sigma)} + 1 = \frac{\beta C k}{\text{trace}(\Sigma)} + 1 = SNR_k,$$

where the notation emphasizes the fact that the SNR depends only on the cardinality.
Therefore, we conclude that

$$\vartheta_j \mid \{\|X\|_0 = k\} = \frac{\text{trace}(\Sigma)(SNR - 1)}{k\beta \|a_j\|^2},$$

that is, if the SNR and the cardinality of the support are given, the scaling is completely
determined. Since we don't know the cardinality, we should treat it as a random
variable and assign some reasonable prior probabilities to it. If we define

$$p_k = P\{\|X\|_0 = k\}, \quad \sum_{k=1}^{n} p_k = 1,$$

the marginal of the scale parameters are

$$\vartheta_j = \sum_{k=1}^{n} p_k \vartheta_j \mid \{\|X\|_0 = k\} = \frac{\text{trace}(\Sigma)(SNR - 1)}{\beta \|a_j\|^2} \sum_{k=1}^{n} \frac{1}{k} p_k. \tag{10.8}$$

Hence, we have found a reasonable expression for the scaling variables. To give a
further interpretation to the formula, let us write (10.8) as

$$\vartheta_j = \frac{\alpha^2}{\|a_j\|^2}, \quad \alpha = \text{constant},$$

and substitute the expression in the formula (10.6), to get

$$\mathscr{E}(x, f(x, \vartheta, 0+)) = \frac{1}{2}\|b - \mathbf{A}x\|^2 + \frac{\sqrt{2}}{\alpha} \sum_{j=1}^{n} \|a_j\| |x_j|.$$

This suggests that in the penalty term, the components should be weighted by the norms of the columns of the forward map $\mathbf{A}$. This type of weighting has been used for a long time in geophysics and in biomedical imaging, and it is known as *sensitivity weighting*. Indeed, in classical signal analysis, the sensitivity of a forward map $x \mapsto f(x)$ to the jth component of x around the value $x = x_0$ is defined as

$$s_j(x_0) = \left\| \frac{\partial f}{\partial x_j}(x_0) \right\|.$$

In the case of a linear forward map, the sensitivity is the same at every point x_0, and is given by

$$s_j = \|a_j\|.$$

The heuristic justification for sensitivity weighting is that to avoid favoring components to which the data are highly sensitive, the penalty should be proportional to the sensitivity. The above argument therefore provides a Bayesian justification for sensitivity weighting in addition to giving a criterion for setting the values of the scale parameters.

10.1.2 IAS with Priorconditioned CGLS

We now turn to the question of how to terminate the iterations. A popular stopping criterion for iterative schemes is to terminate the iteration when the relative change in two consecutive solutions falls below a certain level. In the IAS algorithm, we use this criterion for the hyperparameter θ.

Before incorporating the stopping criterion in the IAS algorithm, consider the updating step of the variable x. The solution of the least squares problem can be approximated in a computationally efficient way by using one of the iterative linear solvers presented in the previous chapter, namely, the priorconditioned CGLS method equipped with a suitable termination rule. We now augment the IAS algorithm with these features.

Iterative Alternating Sequential (IAS) algorithm with priorconditioned CGLS: Given the shape parameter $\beta > 3/2$, an estimate of the SNR, stopping tolerance $\tau > 0$ and maximum number of iterations $t_{\max}$, and data $b \in \mathbb{R}^m$,

1. **Initialize:** Assign the values of the scale parameters ϑ according to (10.8), then set $\theta = \vartheta$, $t = 0$, $\Delta_\theta = \infty$.
2. **Iterate:** While $\Delta_\theta > \tau$ and $t < t_{\max}$,

   (a) Update x using priorconditioned CGLS: Set $\mathsf{A}_\theta = \mathsf{A}\mathsf{D}_\theta^{1/2}$, and compute with CGLS iterates

   $$w_\ell = \mathrm{argmin}\{\|\mathsf{A}_\theta w - b\| \mid x \in \mathcal{K}_\ell(\mathsf{A}_\theta b, \mathsf{A}_\theta^\mathsf{T}\mathsf{A}_\theta)\}.$$

   Terminate the CGLS iteration at the step k, where k is the first index such that either

   $$\|\mathsf{A}_\theta w_{k+1} - b\|^2 < m$$

   or

   $$\|b - \mathsf{A}_\theta w_{k+1}\|^2 + \|w_{k+1}\|^2 > \|b - \mathsf{A}_\theta w_k\|^2 + \|w_k\|^2. \qquad (10.9)$$

   Set $x = \mathsf{D}_\theta^{1/2} w_k$.
   (b) Set $\theta_{\mathrm{old}} = \theta$.
   (c) Update θ, setting

   $$\theta_j = \vartheta_j \left(\frac{\eta}{2} + \sqrt{\frac{\eta^2}{4} + \frac{x_j^2}{2\vartheta_j}} \right), \quad 1 \le j \le n.$$

   (d) Update

   $$\Delta_\theta = \frac{\|\theta - \theta_{\mathrm{old}}\|}{\|\theta_{\mathrm{old}}\|},$$

   and advance the counter, $t \to t + 1$.

Observe that as discussed at the end of the previous chapter, we augmented the stopping criterion of the inner CGLS iteration based on the Morozov discrepancy principle by the condition (10.9) based on the rationale that the CGLS iteration, in fact, seeks to minimize the quadratic expression

$$\left\| \begin{bmatrix} \mathsf{A} \\ \mathsf{D}_\theta^{-1/2} \end{bmatrix} x - \begin{bmatrix} b \\ 0 \end{bmatrix} \right\|^2 = \|b - \mathsf{A}_\theta w\|^2 + \|w\|^2, \quad w = \mathsf{D}_\theta^{-1/2} x,$$

so it is natural to continue the iterations only as long as this quantity keeps decreasing.

10.2 More General Sparse Representations

So far, we have assumed that the unknown vector $x \in \mathbb{R}^n$ itself is a realization of a random variable that is believed to be sparse, or, at least close to sparse. In general, however, we may require that instead of X, a derived quantity,

$$Y = LX, \quad L \in \mathbb{R}^{k \times n},$$

is the sparse signal to be estimated. If $k = n$ and the matrix L is invertible, the inverse problem can be stated in terms of the new variable, but in general, this is not the case. For example, consider a pixel image, modeled as a quadrilateral mesh where the nodes represent the pixel values. Adjacent pixels in the vertical and horizontal directions are connected by edges. For simplicity, we assume here that the pixel values at the boundary nodes vanish, while in the interior nodes the values need to be estimated from some indirect noisy data. If the number of interior nodes is n_v, and the number of edges between the nodes is n_e, we define the increments over the edges as the difference of the nodal values at the endpoints of the edges. The vector of increments y is related to the vector x of interior nodal values through the equation

$$y = Lx, \quad L \in \mathbb{R}^{n_e \times n_v},$$

the matrix L being sparse and containing at most a pair $(+1, -1)$ of nonzero values in each row. Considering only the pixel values at the interior nodes, we see that the null space of L is trivial,

$$\mathcal{N}(L) = \{0\}. \tag{10.10}$$

Indeed, if all increments vanish so that $Lx = 0$, connecting any interior node to a boundary node by a chain of edges shows that the value at that node must vanish, and therefore, we may conclude that $x = 0$.

To characterize the range of L, consider the circulation around any elementary loop T_k in the network, see Fig. 10.2. For consistency, the signed sum of the increments

Fig. 10.2 The compatibility condition expresses the fact that the circulation around each closed loop in the network must vanish. At the boundary, since the function is zero at boundary nodes, the loop comprises only the free edges, that is, edges with at least one end node that is not a boundary node

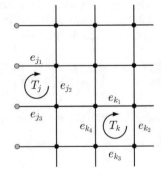

over the edges forming the loop must vanish. Therefore, if n_t is the number of elementary loops, there is a matrix $\mathsf{M} \in \mathbb{R}^{n_t \times n_e}$ whose elements are zeros and ± 1s, such that

$$\mathsf{M}y = 0. \tag{10.11}$$

It is not difficult to deduce that condition (10.11) holds exactly for increment vectors corresponding to some nodal values $y = \mathsf{L}x$, that is,

$$\mathcal{R}(\mathsf{L}) = \mathcal{N}(\mathsf{M}). \tag{10.12}$$

The conditions (10.10) and (10.12) can be stated by saying that L and M form a *short exact chain*, and expressed formally as

$$\{0\} \longrightarrow \mathbb{R}^{n_v} \xrightarrow{\mathsf{L}} \mathbb{R}^{n_e} \xrightarrow{\mathsf{M}} \mathbb{R}^{n_t} \longrightarrow \{0\}.$$

Condition (10.11) is a *compatibility condition* that the vector y needs to satisfy.

Assume now that we believe, a priori, that the image has a sparse representation, in the sense that there are only a few nonzero increments, or at least, few increments above some small threshold value. We write the corresponding conditionally Gaussian prior for the random variable Y as

$$\pi_{Y|\Theta}(y \mid \theta) \propto \exp\left(-\frac{1}{2} \sum_{j=1}^{n_e} \frac{y_j^2}{\theta_j} - \frac{1}{2} \sum_{j=1}^{ne} \log \theta_j \right).$$

However, since the data, hence the likelihood, is expressed in terms of X, we would like to express the prior in terms of X rather than Y. Formally, the dimensionality of Y is much higher than that of X, however, the compatibility condition (10.11) forces Y into a lower dimensional subspace, the null space of M, which is of the same dimension as X. The problem that arises is how to restrict the prior density to the subspace $\mathcal{N}(\mathsf{M})$. Fortunately, numerical linear algebra comes to our rescue here. Rather than building the connection through the matrix M, we show how to find a computationally feasible strategy to modify the problem while solving it.

We start by introducing the auxiliary variable

$$z = \mathsf{D}_\theta^{-1/2} y, \quad \mathsf{D}_\theta = \begin{bmatrix} \theta_1 & & \\ & \ddots & \\ & & \theta_{n_e} \end{bmatrix}$$

and express the compatibility condition $y \in \mathcal{N}(\mathsf{M}) = \mathcal{R}(\mathsf{L})$ as

$$z \in \mathcal{R}(\mathsf{L}_\theta), \quad \mathsf{L}_\theta = \mathsf{D}_\theta^{-1/2}\mathsf{L}. \tag{10.13}$$

The matrix L_θ has more rows than columns, therefore its QR-decomposition is of the form

$$L_\theta = QR = \begin{bmatrix} Q_1 & Q_2 \end{bmatrix} \begin{bmatrix} R_1 \\ O \end{bmatrix},$$

where the matrix $R_1 \in \mathbb{R}^{n_v \times n_v}$ is upper triangular with nonzero diagonal entries, because from (10.10) it follows that L and L_θ have rank n_v. Condition (10.13) guarantees that there is $x \in \mathbb{R}^{n_v}$ such that

$$z = L_\theta x = QRx.$$

Multiplying both sides by Q^T, and using the fact that $Q^\mathsf{T}Q = I$, it follows that

$$Q^\mathsf{T} z = \begin{bmatrix} Q_1^\mathsf{T} z \\ Q_2^\mathsf{T} z \end{bmatrix} = Rx = \begin{bmatrix} R_1 x \\ 0 \end{bmatrix}. \tag{10.14}$$

Therefore, any vector z in the range of L_θ satisfies the compatibility condition

$$Q_2^\mathsf{T} z = 0, \quad \text{or} \quad z \in \mathscr{H} = \mathcal{N}(Q_2^\mathsf{T}).$$

We include this condition as part of the prior for the random variable $Z = D_\theta^{-1/2} Y$, conditioned on $\Theta = \theta$ by setting

$$\pi_{Z|\Theta}(z \mid \theta) \propto \exp\left(-\frac{1}{2} \sum_{j=1}^{n_e} z_j^2 - \frac{1}{2} \sum_{j=1}^{n_e} \log \theta_j \right) \delta_{\mathscr{H}}(z),$$

where $\delta_{\mathscr{H}}(z)$ is a density that accumulates all the probability mass on the subspace $\mathscr{H}$. Moreover, we can use (10.14) to write the likelihood in terms of z as

$$\pi_{B|Z}(b \mid z) \propto \exp\left(-\frac{1}{2} \| b - AR_1^{-1} Q_1^\mathsf{T} z \|^2 \right).$$

According to Bayes' formula, the posterior density for the pair (Z, Θ) is

$$\pi_{Z,\Theta|B}(z, \theta)$$
$$\propto \exp\left(-\frac{1}{2} \| b - AR_1^{-1} Q_1^\mathsf{T} z \|^2 - \frac{1}{2} \sum_{j=1}^{n_e} z_j^2 - \frac{1}{2} \sum_{j=1}^{n_e} \log \theta_j \right) \delta_{\mathscr{H}}(z) \pi_\Theta(\theta),$$

and the MAP estimate can be computed with the IAS algorithm, where the updating of z is done by solving the least squares problem

$$z = \operatorname{argmin} \left\| \begin{bmatrix} AR_1^{-1} Q_1^\mathsf{T} \\ I_{n_e} \end{bmatrix} z - \begin{bmatrix} b \\ 0 \end{bmatrix} \right\|^2.$$

The last point that needs to be addressed is how to ensure that the IAS algorithm enforces the compatibility condition $z \in \mathcal{H}$. Fortunately, the compatibility condition is automatically obeyed by the IAS iterates. To verify it, write

$$z = z_1 + z_2, \quad z_1 \in \mathcal{H}, \quad z_2 \in \mathcal{H}^\perp,$$

and observe that from

$$\mathsf{Q}_1^\mathsf{T} z_2 = 0,$$

it follows that

$$\left\| \begin{bmatrix} \mathsf{AR}_1^{-1}\mathsf{Q}_1^\mathsf{T} \\ \mathsf{I}_{n_e} \end{bmatrix} z - \begin{bmatrix} b \\ 0 \end{bmatrix} \right\|^2 = \|b - \mathsf{AR}_1^{-1}\mathsf{Q}_1^\mathsf{T} z\|^2 + \|z\|^2$$

$$= \|b - \mathsf{AR}_1^{-1}\mathsf{Q}_1^\mathsf{T} z_1\|^2 + \|z_1\|^2 + \|z_2\|^2.$$

Therefore the least squares solution always satisfies $z_2 = 0$, because any nonzero component of z_2 would make the sum of the squares larger.

We conclude by pointing out that since

$$\mathsf{L}_\theta^\dagger = \mathsf{R}_1^{-1}\mathsf{Q}_1^\mathsf{T},$$

we may organize the steps of the IAS algorithm as follows.

IAS algorithm with priorconditioned CGLS, generalized sparsity condition: Given the shape parameter $\beta > 3/2$, an estimate of the SNR, stopping tolerance $\tau > 0$, maximum number of iterations $t_{\max}$, and data $b \in \mathbb{R}^m$,

1. **Initialize:** Compute the scale parameters ϑ by (10.8). Set $\theta = \vartheta, t = 0, \Delta_\theta = \infty$.
2. **Iterate:** While $\Delta_\theta > \tau$ and $t < t_{\max}$,

   (a) Update x using priorconditioned CGLS: Set $\mathsf{A}_\theta = \mathsf{AL}_\theta^\dagger$, with $\mathsf{L}_\theta = \mathsf{D}_\theta^{-1}\mathsf{L}$, and compute the CGLS iterates

   $$z_\ell = \operatorname{argmin}\{\|\mathsf{A}_\theta z - b\| \mid x \in \mathcal{K}_\ell(\mathsf{A}_\theta b, \mathsf{A}_\theta^\mathsf{T} \mathsf{A}_\theta)\}.$$

   Terminate the CGLS iteration at the step k, where k is the first index such that either

   $$\|\mathsf{A}_\theta z_k - b\|^2 < m$$

   or

   $$\|b - \mathsf{A}_\theta z_{k+1}\|^2 + \|z_{k+1}\|^2 > \|b - \mathsf{A}_\theta z_k\|^2 + \|z_k\|^2.$$

   Set $x = \mathsf{L}_\theta^\dagger z_k$.
   (b) Set $\theta_{\text{old}} = \theta$.

(c) Given the updated z, update θ, evaluating

$$\theta_j = \vartheta_j \left(\frac{\eta}{2} + \sqrt{\frac{\eta^2}{4} + \frac{z_j^2}{2\vartheta_j}} \right), \quad 1 \le j \le n_e.$$

(d) Update

$$\Delta_\theta = \frac{\|\theta - \theta_{\text{old}}\|}{\|\theta_{\text{old}}\|},$$

and advance the counter, $t \to t + 1$.

Before presenting some computed examples, a comment on the implementation of the CGLS in this setting is in order. The action of the matrix $\mathsf{AL}_\theta^\dagger$ on a given vector u can be computed in two steps: First, solve in the least squares sense the problem

$$\mathsf{L}_\theta v = u,$$

then multiply v by A. To compute the product of a vector with the transpose of $\mathsf{AL}_\theta^\dagger$, recall that

$$\left(\mathsf{AL}_\theta^\dagger \right)^\mathsf{T} = \mathsf{L}_\theta \left(\mathsf{L}_\theta^\mathsf{T} \mathsf{L}_\theta \right)^{-1} \mathsf{A}^\mathsf{T}.$$

To take advantage of the sparsity of the matrix $\mathsf{L}_\theta^\mathsf{T} \mathsf{L}_\theta$, first solve the linear system

$$\left(\mathsf{L}_\theta^\mathsf{T} \mathsf{L}_\theta \right) w = \mathsf{A}^\mathsf{T} v,$$

for w, then multiply w by L_θ to get

$$\mathsf{L}_\theta w = \mathsf{L}_\theta \left(\mathsf{L}_\theta^\mathsf{T} \mathsf{L}_\theta \right)^{-1} \mathsf{A}^\mathsf{T} v = \left(\mathsf{AL}_\theta^\dagger \right)^\mathsf{T} v.$$

We close this chapter with computed examples demonstrating the viability of the method.

10.3   Some Examples

The following three examples elucidate the different aspects of the IAS algorithm. We start with a rather elementary one-dimensional inverse problem and then move to two-dimensional problems where the sensitivity and dimensionality of the data play a role.

Example 10.1   In this example, we consider the problem of estimating a function $u : [0, 1] \to \mathbb{R}$ from noisy observations of a convolution of it,

$$g(t) = \int_0^1 a(t-s)u(s)ds + \text{noise},$$

where a is a given convolution kernel. We discretize the problem as follows: Divide the interval $[0, 1]$ into n subintervals of length $h = 1/n$ by discretization points

$$s_j = \frac{j}{n}, \quad 0 \le j \le n,$$

and write an approximation

$$\int_0^1 a(t-s)u(s)ds \approx h \sum_{j=1}^n a(t-s_j)u(s_j) = h \sum_{j=1}^n a(t-s_j)z_j.$$

The data are assumed to be observed at some of the discretization points t_k,

$$b_k = g(t_k) = h \sum_{j=1}^n a(t_k - s_j)z_j + e_k,$$

or, in matrix form,

$$b = \mathsf{A}z + e,$$

where we assume that e is a realization of a Gaussian random variable. We assume that while the signal z itself is not sparse, the vector x of its increments is, where

$$x_j = z_j - z_{j-1} = j\text{th increment of } z.$$

Hence, we write the sparsity promoting prior for the increment vector. To this end, we need to express the likelihood, too, in terms of the increments. Assuming that $z_0 = 0$, we have the telescoping sum representation,

$$z_k = z_k - z_{k-1} + (z_{k-1} - z_{k-2}) + \ldots + (z_1 - z_0) = \sum_{j=1}^k x_j, \qquad (10.15)$$

which allows to write the matrix transformations between x and z as

$$x = \mathsf{L}z, \quad \mathsf{L} = \begin{bmatrix} 1 & & & \\ -1 & 1 & & \\ & \ddots & \ddots & \\ & & -1 & 1 \end{bmatrix},$$

and

$$z = \mathsf{L}^{-1}x, \quad \mathsf{L}^{-1} = \begin{bmatrix} 1 & & & \\ 1 & 1 & & \\ \vdots & & \ddots & \\ 1 & 1 \dots & & 1 \end{bmatrix},$$

The form of the inverse matrix can be deduced from formula (10.15). In terms of x, the forward model is

$$b = \mathsf{A}\mathsf{L}^{-1}x + e = \mathsf{A}_\mathsf{L}x + e, \quad \mathsf{L}z = x,$$

where

$$\mathsf{A}_\mathsf{L} = \mathsf{A}\mathsf{L}^{-1}.$$

We can now write the Gibbs energy for the increment vector using the hierarchical prior, assuming that the Gaussian noise has been whitened,

$$\mathscr{E}(x, \theta) = \frac{1}{2}\|b - \mathsf{A}_\mathsf{L}x\|^2 + \frac{1}{2}\sum_{j=1}^{n}\frac{x_j^2}{\theta_j} - \eta\sum_{j=1}^{n}\log\theta_j + \sum_{j=1}^{n}\frac{\theta_j}{\vartheta_j}.$$

To generate the data, the unit interval is divided into 128 equal subintervals, and we define the generative model as a piecewise constant function with five discontinuities, Thus, $\|x\|_0 = 5$. The kernel a is chosen to be a Gaussian, such that after discretization,

$$a_{jk} = 0.05\, e^{-|s_j - t_k|^2/2\lambda^2}, \quad \lambda = 0.02.$$

To demonstrate the effect of the parameters, we start by choosing the shape and scaling parameters as

$$\eta = \beta - \frac{3}{2} = 0.1, \quad \vartheta_j = \vartheta_0 = 0.1.$$

The stopping tolerance for the outer iterations is set at $\tau = 10^{-3}$, and in the inner iteration, the least squares problem for updating x is solved by using the standard system solver (mldivide, or "backslash"), and the algorithm converged in 60 outer iterations. Figure 10.3, first row, shows the outcome. We observe that η is not small enough to promote the sparsity of the increment, and the estimated z is more like a smooth function than a piecewise constant one as we hope. We observe that the dynamical range of the estimate, controlled mostly by the parameter ϑ_0, corresponds well to that of the generative model. To improve the quality of the estimate, we decrease significantly the parameter η that controls the sparsity, choosing

$$\eta = \beta - \frac{3}{2} = 10^{-3}, \quad \vartheta_j = \vartheta_0 = 0.1.$$

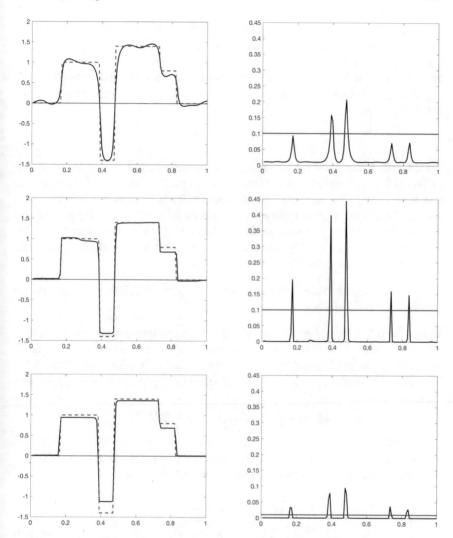

Fig. 10.3 Results of the IAS algorithm, with parameter values from top to bottom as follows: $(\eta, \vartheta_0) = (0.1, 0.1)$ (top row), $(\eta, \vartheta_0) = (10^{-3}, 0.1)$ (middle row), and $(\eta, \vartheta_0) = (10^{-3}, 10^{-2})$ (bottom row). On the left, the generative model is plotted as a dashed red curve. On the right, the value ϑ_0 is indicated by the horizontal red line. The results corroborate the theoretical argument that η controls the sparsity, and ϑ_0 the dynamical range, and together with η, the cleanliness of the estimate outside the singularities

The results are shown in the second row of Fig. 10.3. This time, the solution is close to a piecewise constant function, and θ is in general small, with a few significant outliers at and around the points of discontinuities that match well with the ones of the generative model. The iterations converged in 165 outer iteration rounds.

Finally, to test the effect of the scaling parameter, we run the algorithm with parameter values

$$\eta = \beta - \frac{3}{2} = 10^{-3}, \quad \vartheta_j = \vartheta_0 = 10^{-2}.$$

The IAS algorithm converged after 153 outer iterations, and the results are shown in the third row of Fig. 10.3. We observe two things: First, the size of the estimated θ is significantly smaller, due to the decreased value of ϑ_0, which in turn translates into an insufficient dynamical range of the vector z. Moreover, since the product $\vartheta_0 \eta$ controls the size of θ_j at points outside the support of the increment vector, as this product is smaller by a factor of ten, the estimate is flatter between the discontinuities corresponding to the ones of the generative model.

The following example demonstrates the use of the sensitivity scaling and the CGLS in the IAS algorithm.

Example 10.2 We consider a linear inverse source problem in two dimensions, having certain similarities with the inverse problems of electroencephalography (EEG) and magnetoencephalography (MEG). In our example, consider the unit disc Ω in $\mathbb{R}^2$ containing a distribution $\rho(r), r \in \Omega$ of sources that generate a field $b(r')$ outside Ω, such that

$$b(r') = \int_\Omega \frac{\rho(r)}{|r - r'|^2} dr, \quad r' \in \mathbb{R}^2 \setminus \overline{\Omega}.$$

We assume that the field b is measured at points $r'_j, 1 \leq j \leq m$. To set up the discrete forward model, we approximate the integral by a quadrature,

$$\int_\Omega \frac{\rho(r)}{|r - r'|^2} dr \approx \sum_{j=1}^{n} w_j \frac{\rho(r_j)}{|r_j - r'|^2},$$

where w_js are the quadrature weights and r_js are the corresponding quadrature nodes. We denote the unknowns by $q_j = w_j \rho(r_j)$, and write the observation model as

$$b_k = \sum_{j=1}^{n} \frac{q_j}{|r_j - r'_k|^2} + e_k, \quad 1 \leq k \leq m,$$

where e_k is the additive noise. The discretization and the measurement geometry are shown on the left of Fig. 10.4. On the right, the generative model with three-point sources is shown, together with the data, contaminated by additive Gaussian independent error with standard deviation $\sigma = 0.05$, corresponding to 1.1% of the maximum amplitude of the data, or 2.5% of the mean amplitude.

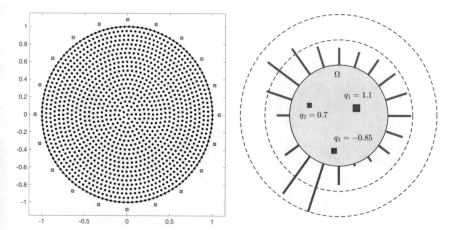

Fig. 10.4 Left: Quadrature points used to discretize the forward model ($n = 1\,301$, the green squares indicating the measurement points ($m = 20$). Right: The generative model with three-point sources in Ω (shaded region), and visualization of the data. Positive sources are marked by red and negative by blue. The same color coding applies to the data

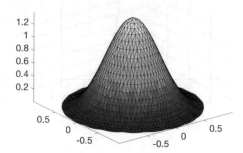

Fig. 10.5 Computed values of ϑ_j using the formula (10.8), linearly interpolated between the nodal values. The lowest sensitivity at the center of the disc leads to the highest value of ϑ_j

To run the IAS algorithm, we initialize the process by computing the scaling parameter ϑ using the sensitivity argument and formula (10.8). We assume a priori that the source consists of at most 10 active sources, with no preference for any cardinality below that limit, thus setting

$$p_k = \frac{1}{10} \text{ for } 1 \le k \le 10, \text{ and } p_k = 0 \text{ for } k > 10,$$

implying that

$$\sum_{k=1}^{n} \frac{1}{k} p_k = \frac{1}{10} \sum_{k=1}^{10} \frac{1}{k} \approx 0.293.$$

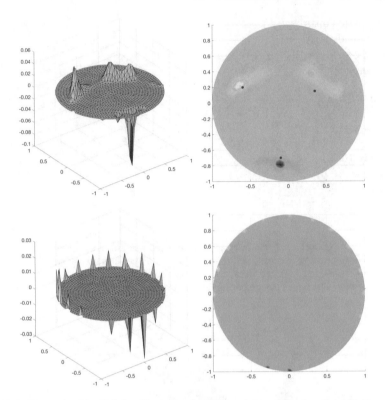

Fig. 10.6 The result of the IAS algorithm using the SNR-exchangeability-based sensitivity scaling (top row) and the corresponding result obtained by setting $\vartheta_j = \vartheta_0$ = constant. In the former case, the algorithm converged in 101 outer iterations, while the latter case required 38 iterations. The positions of the sources of the generative model are indicated by red dots in the upper row

We visualize the distribution of the values of ϑ_j by plotting a piecewise linear surface over Ω interpolating between the values ϑ_j at the nodal points r_j, shown in Fig. 10.5. As expected, the sensitivity in the center of the disc is lowest, leading to the largest value of ϑ_j, and at nodal points closest to the observation points r'_k it is the highest.

We run the IAS algorithm using the priorconditioned CGLS updating strategy, setting the stopping criterion for the outer iteration at $\tau = 10^{-3}$. The stopping criterion was reached after 108 iterations. Figure 10.6 shows the result in the top row. For comparison, we run the algorithm without the sensitivity weighting, setting $\vartheta_j = \vartheta_0$, a constant value throughout. The constant was chosen so that it corresponds to ϑ_j at the center of the disc when sensitivity is taken into account, $\vartheta \approx 1.38$. The results in the bottom row of Fig. 10.6 show a complete lack of depth resolution: The algorithm explains the data by adjusting the source at the discretization points nearest to the receivers, having the highest sensitivity.

Finally, we point out that the dynamic range of the solution does not correspond to the source values of the generative models, which is due to the fact that the reconstruction distributes the source to several source locations as well as to the fact that the quadrature weights are lumped with the source density. For an analysis of the source strengths, an integral of the sources should be computed to compare with the generative values.

The third example demonstrates how the IAS algorithm works when the mapping from the primary unknown and its sparse representation is not one-to-one. We consider a classical X-ray tomography problem with limited projection data.

Example 10.3 Let $\Omega \subset \mathbb{R}^2$ denote the object that is illuminated by X-ray sources. The goal is to estimate the density distribution inside Ω from the transmission attenuation of the X-rays through it. Denoting by $\rho(r)$ the density, we assume that the X-ray attenuation of the initial intensity I_0 is described by the Beer–Lambert law stating that the attenuation of the intensity between the source at r_{source} and the receiver at r_{receiver} along a line ℓ of length L and parametrized by the arclength,

$$\ell : r(s) = (1 - s/L)r_{\text{source}} + s/L \, r_{\text{receiver}}, \quad 0 \leq s \leq L,$$

is given by

$$I = I_0 \exp\left(-\int_0^L \rho(r(s)) ds\right).$$

Assuming that the object is illuminated from different directions, the attenuation being measured over m lines $\ell^{(k)}$, $1 \leq k \leq m$ the noiseless data can be described by the linear model

$$b_k = \int_0^{L_k} \rho(r^{(k)}(s)) ds, \quad 1 \leq k \leq m, \tag{10.16}$$

where $r^{(k)}(s)$ is the parametrization of the kth ray, and L_k is its length. In this example, we assume that Ω is a unit disc, and for discretization, it is approximated by a polygon divided in small, approximately equilateral triangles. Denoting by r_j the jth vertex in the triangular mesh, we discretize ρ for the forward problem by denoting

$$x_j = \rho(r_j),$$

and write

$$\rho(r) = \sum_{j=1}^n x_j \psi_j(r),$$

where ψ_j is a piecewise linear Lagrange basis function such that

$$\psi_j(r_\ell) = 1, \text{ if } j = \ell, \text{ and } 0 \text{ otherwise.}$$

Then,

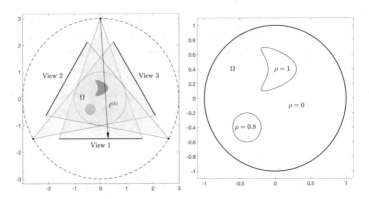

Fig. 10.7 The X-ray tomography arrangement with three views (left), and the target used to generate the test data. Each view angle corresponds to 400 lines of integration across the target

$$\int_0^{L_k} \rho\big(r^{(k)}(s)\big)ds \approx \sum_{j=1}^n x_j \int_0^{L_k} \psi_j\big(r^{(k)}(s)\big)ds = \sum_{j=1}^n a_{kj}x_j.$$

Taking into account the observation error and modeling errors, we arrive at the linear model

$$b = Ax + \varepsilon, \quad A \in \mathbb{R}^{m \times n}.$$

We will model the error ε as Gaussian noise, which is more a choice of convenience than a result of analysis of the sources of error.

In the computed example, we assume that the target Ω is illuminated from few directions by a fan beam source, and the attenuation is measured on the opposite side of the object along a line segment, see Fig. 10.7. Moreover, the prior belief is that the target density is a piecewise constant function, implying that the increments over the edges joining the nodes in the triangular mesh comprise a sparse vector. Thus, if n_e is the number of edges, and the edges are defined in terms of the end point nodes, $e_\ell = \{r_{i_\ell}, r_{j_\ell}\}$, we define a matrix $L \in \mathbb{R}^{n_e \times n}$ such that

$$\big(Lx\big)_\ell = x_{i_\ell} - x_{j_\ell}.$$

The edges outnumber the nodes as Euler's formula for planar graphs implies, so the matrix L has more rows than columns and cannot be invertible. Therefore, we need to apply the ideas outlined in Sect. 10.2.

We start by generating the data, considering two data sets; in one we use five illumination angles, and in the other, 15 illumination angles are used, in both cases, uniformly distributed around the circle. Each illumination angle corresponds to 400 lines of integration over the opening angle with a full view of Ω as indicated in the figure. The generative model used for the data generation is shown in Fig. 10.7. We add Gaussian independent and equally distributed noise to the noiseless data, with

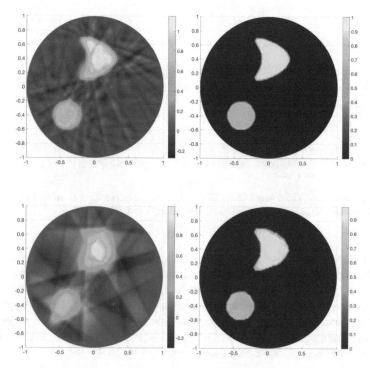

Fig. 10.8 Top row: estimate of the density based on 15 uniformly distributed projections. The estimate on the left is computed by the CGLS algorithm with early stopping at the discrepancy, and the one on the right with the IAS iteration. Bottom row: The data consist of five projection directions

standard deviation corresponding to 0.1% of the maximum attenuation over all lines. The area Ω is divided into triangles with a diameter of the order 0.02, computed with a standard mesh generator (see Notes and Comments) yielding a tessellation of 9 062 nodes, 26 874 edges, and 17 813 triangles.

To initialize the IAS algorithm, we set $\vartheta_j = \vartheta_0$, that is, we do not use the sensitivity scaling, and choose the constant $\vartheta_0 = 0.1$, large enough to allow the variances θ_j to allow jumps of the order of magnitude ~ 1. To keep the background clean, we choose η so that the product $\eta\vartheta_0$ is close to zero, setting $\eta = 10^{-5}$. This choice is also favoring a sparse solution. The maximum allowed number of outer IAS iterations is limited to 50, with the tolerance $\tau = 10^{-2}$.

Consider first the case with more complete data, assuming 15 projection directions. In this case, the dimension of the data is $m = 15 \times 400 = 6\,000$. The requested tolerance τ is reached in 21 IAS iteration rounds. It is of interest to see how many inner iterations of the priorconditioned CGLS are required . The numbers are shown in Fig. 10.9, indicating the efficiency of the algorithm: The first few outer iterations require more work, and as the algorithm proceeds, only a few inner iterations are needed. In Fig. 10.8, the left column shows the reconstruction using plain CGLS

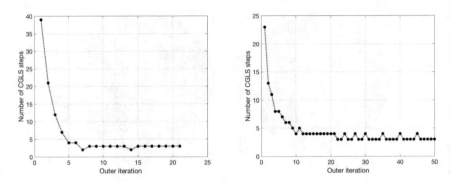

Fig. 10.9 Number of priorconditioned CGLS iterations in each outer iteration of the IAS algorithm. The left panel corresponds to 15 projections, and the right to 5 projections

algorithm with an early stopping at the discrepancy as a regularization, and on the right, the outcome of the IAS algorithm. The former results contain the characteristic streak artifacts typical for reconstructions from a small number of projections, while in the IAS reconstruction, the background as well as the constant density objects are free of artifacts. For comparison, we run the same algorithms with only 5 projections. The corresponding results are shown in the second row of Fig. 10.8, and the numbers of inner iterations are shown in the right panel of Fig. 10.9.

Notes and Comments

The IAS algorithm as presented here has been developed and discussed in a series of articles [9, 11, 12, 19, 22], with a particular application to functional brain imaging by magnetoencephalography. The hypermodel for the variance Θ need not be a gamma distribution, and in fact, hypermodels from a wider class of generalized gamma distributions were considered, however, the convexity of the objective function guaranteeing uniqueness is no longer warranted. The ideas of using non-convex objective functions are developed further in the references [14, 15, 61].

The triangulation of the computational domain in the example 9.3 was done by using the mesh generator DistMesh, described in detail in the article [60].

Chapter 11
Sampling: The Real Thing

In Chap. 4, where we first introduced random sampling, we pointed out that sampling is used primarily to explore a given probability distribution and to calculate estimates of integrals via Monte Carlo integration. It was also indicated that sampling from a non-Gaussian probability density may be a challenging task. In this section we further develop the topic and introduce Markov chain Monte Carlo (MCMC) sampling.

11.1 Preliminaries: Markov Chains and Random Walks

In this section, we introduce the basic concepts of random walks, transition matrices and kernels, Markov chains, and their connections to sampling. We start with a discrete model example.

11.1.1 An Introductory Example

As an introduction to sampling by using Markov chains, we start with a motivational example in discrete state space. Let S be a finite state space,

$$S = \{1, 2, 3, 4, 5\},$$

and let X be a random variable with possible values in S, that is, $X : \Omega \to S$. We denote the probabilities of different values by

$$p_j = \mathrm{P}\{X = j\}, \quad 1 \le j \le 5, \quad \sum_{j=1}^{5} p_j = 1,$$

© The Author(s), under exclusive license to Springer Nature Switzerland AG 2023
D. Calvetti and E. Somersalo, *Bayesian Scientific Computing*, Applied Mathematical
Sciences 215, https://doi.org/10.1007/978-3-031-23824-6_11

and introduce the corresponding probability vector

$$\pi_X = \begin{bmatrix} p_1 \\ \vdots \\ p_5 \end{bmatrix}.$$

In Sect. 4.4, we constructed an algorithm for generating independently drawn realizations from a discrete probability distribution. With that algorithm, we can compute a sample of size T of independently drawn realizations,

$$\mathscr{S} = \{x^1, x^2, \ldots, x^T\}, \quad x^t \in S,$$

with the property that asymptotically, as T increases, the relative frequencies of possible output values converge to the probability of the value, that is,

$$\nu_j^T = \frac{\#\{x^t = j\}}{T} \to p_j.$$

When the state space is small as in the present example, the independent sampling approach is effective and easy to implement, but as seen in Chap. 4, when the state space dimension increases, alternative ways are needed. In particular, we are interested in a sequential sampling approach, in which the next realization x^{t+1} can be computed from the current realization x^t alone.

In order to define a sequential algorithm, let x^t denote the current realization, interpreted as the position of a random walker at time t in the state space, and the next realization x^{t+1} is obtained by choosing a transition $x^t \to x^{t+1}$ in a stochastic manner. To that end, we define a *transition matrix* $\mathsf{P} \in \mathbb{R}^{5 \times 5}$ with entries

$$p_{jk} = \text{probability of moving from node } k \text{ to node } j.$$

In particular, P is defined so that the matrix entries are all non-negative, and the sum of the entries in each column is one. To give a concrete example of a transition matrix, consider the directed network in Fig. 11.1, where the nodes correspond to the state space S and the directed links correspond to possible transitions. In this example, we assume that if at time t the random walker is in node k, a move to another node j is allowed if and only if there is a directed link $k \to j$. Furthermore, for the sake of simplicity, we postulate that all possible moves from node k have equal probability. In the case of the network of Fig. 11.1, this leads to a transition matrix given by

$$\mathsf{P} = \begin{bmatrix} 0 & 1 & 0 & 1/3 & 0 \\ 1/2 & 0 & 0 & 1/3 & 0 \\ 0 & 0 & 0 & 0 & 1/2 \\ 1/2 & 0 & 1/2 & 0 & 1/2 \\ 0 & 0 & 1/2 & 1/3 & 0 \end{bmatrix}, \tag{11.1}$$

Fig. 11.1 A simple directed network with five nodes. The outlinks from node 4 are shown in red, and the moves out of node 4 have all the same probability $1/3$

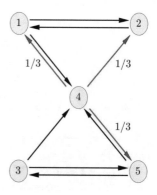

where the probabilities of the moves from the current position are listed in the corresponding column of the transition matrix. Having the transition matrix P, we now define a random walk over the state space as follows:

1. Initialize by choosing one node, e.g., $x^1 = 1$. Set $t = 1$, and define T as the maximum number of steps.
2. While $t < T$, repeat:

   (a) Draw a node ℓ, $1 \le \ell \le 5$ using probabilities in the column x^t of the matrix P.
   (b) Set $x^{t+1} = \ell$, and advance the counter $t \to t + 1$.

The process above generates a random sample $\{x^1, x^2, \ldots, x^T\}$ in a sequential manner, however, in order for the sample to indeed represent realizations of the random variable X that we are interested in, the matrix P must be carefully tailored. To see how the generated sample values are distributed when P is given, and under which conditions this distribution coincides with that of X, we introduce some tools to analyze the sample, using the matrix P of (11.1) as an example.

We define a sequence of random variables X^t, $t = 1, 2, \ldots$, with values in the state space S such that X^{t+1} indicates the position of the random walk after t steps. If we start the random walk from node 1, the probability vector of X^2, whose entries are the probabilities of possible outcomes of the move, is given by the first column of P, that is,

$$\pi_{X^2 | X^1 = 1} = P e_1 = \begin{bmatrix} 0 \\ 1/2 \\ 0 \\ 1/2 \\ 0 \end{bmatrix}. \tag{11.2}$$

Observe that in this case the vector e_1 represents the probability vector of X^1 when we know for sure that the walk starts from node 1. What is the probability vector of X^3, given that the walk started at node 1? Since by (11.2), X^2 has two equiprobable realizations, $x^2 = 2$ and $x^2 = 4$, we conclude that

$$\pi_{X^3|X^1=1} = \frac{1}{2}Pe_2 + \frac{1}{2}Pe_4 = P\begin{bmatrix} 0 \\ 1/2 \\ 0 \\ 1/2 \\ 0 \end{bmatrix} = P^2 e_1.$$

Inductively, the probability vector after t steps is

$$\pi_{X^{t+1}|X^1=1} = P\pi_{X^t|X^1=1} = P^2\pi_{X^{t-1}|X^1=1} = \cdots = P^t e_1.$$

So far, we assumed that the random walk started from node 1. More generally, we may ask what is the probability vector of X^{t+1} if the initial state is chosen randomly. If we assign equal probability $1/5$ to every initial state, then the marginal probability vector is

$$\pi_{X^{t+1}} = \sum_{j=1}^{5} \pi_{X^{t+1}|X^1=j} P\{X^1 = j\} = P^t u, \quad u = \frac{1}{5}\begin{bmatrix} 1 \\ 1 \\ 1 \\ 1 \\ 1 \end{bmatrix}. \tag{11.3}$$

Here, u is the probability vector of the uniform distribution.

A remarkable property of the stochastic process outlined above is that it forgets the past in the following sense: Suppose that you know the state X^t, that is, $X^t = x^t$ for some $x^t \in \{1, 2, 3, 4, 5\}$. The probability vector of the next state, regardless of how one arrived at the current node, is always

$$\pi_{X^{t+1}|X^t=x^t} = Pe_j, \quad x^t = j \text{ for some } j,$$

that is, if the path to $X^t = x^t$ is $X^1 = x^1, X^2 = x^2, \ldots, X^{t-1} = x^{t-1}$, we have

$$\pi_{X^{t+1}|X^1=x^1,\ldots,X^t=x^t} = \pi_{X^{t+1}|X^t=x^t}. \tag{11.4}$$

We say that a *discrete time stochastic process* $\{X^1, X^2, \ldots\}$ is a *Markov process* if it has the property (11.4). This condition is often expressed by saying that "tomorrow depends on the past only through today."

We have now enough tools to analyze the relation between the original question of generating a sample from a given distribution and generating a sample by using a transition matrix. Assume that a transition matrix P, not necessarily the one of our example, is given, defining a Markov chain $\{X^1, X^2, \ldots\}$. The probability distributions of consecutive random variables X^t and X^{t+1} are related through

$$\pi_{X^{t+1}} = P\pi_{X^t}.$$

Fig. 11.2 The visiting histogram with different values of T. The horizontal black lines indicate the sizes of the components of the eigenvector of the transition matrix P corresponding to the eigenvalue $\lambda = 1$, normalized so that the sum of the entries is one

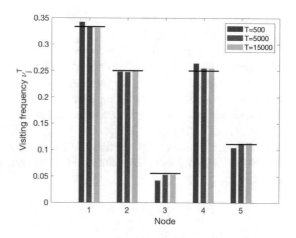

To guarantee that the resulting sample represents draws from some distribution with probability vector π_X, we require that every X^t is distributed according to π_X, which implies, in particular, that

$$\pi_X = P\pi_X. \tag{11.5}$$

This is equivalent to saying that π_X is an eigenvector of the transition matrix P with corresponding eigenvalue $\lambda = 1$. The above condition is often expressed by saying that π_X is an invariant probability vector of the transition matrix P. Let us explore empirically what the probability vector π_X would look like for the matrix (11.1). Given P, we generate a sample $\{x^1, x^2, \ldots, x^T\}$ and compute the relative visiting frequencies of each node. In Fig. 11.2, the relative frequencies of the visits in different nodes with sample sizes $T = 500$, $T = 5\,000$ and $T = 15\,000$, respectively, are plotted as histograms. On the other hand, it turns out that the matrix P has indeed $\lambda = 1$ as an eigenvalue, and that the corresponding eigenvector can be scaled so that its components are all positive and sum up to one, hence can be interpreted as a probability vector. The sizes of the components of this eigenvector are also indicated in Fig. 11.2, confirming that the sampling-based relative frequencies indeed converge toward these values.

We close the discussion of this example with some comments about the underlying theory. It can be shown that a matrix with non-negative entries and column sums equal to one has $\lambda = 1$ as an eigenvalue. Moreover, if the entries are strictly positive, the Frobenius–Perron Theorem (see Notes and Comments for reference) ensures that $\lambda = 1$ is the eigenvalue of largest modulus with geometric multiplicity one, and that a corresponding eigenvector can be scaled so that it is a probability vector. Finally, one can show that regardless of the initial state, the probability vectors

$$\pi_{X^{t+1}} = P^t \pi_{X^1}$$

converge to this unique eigenvector, which is the core idea of the power method to compute the eigenvector. Therefore, regardless of what the random draws of x^t are at the beginning, asymptotically their probability vectors are closer and closer to the limit vector. This is the core idea of Markov chain Monte Carlo (MCMC) sampling.

In the discussion above, we assumed that the transition matrix P was given, and found the associated invariant probability vector π_X as an eigenvector of P. Conversely, suppose that π_X is known instead and that we want to generate a sample distributed according to π_X. The problem therefore is how to find a transition matrix P such that condition (11.5) is satisfied. Answering this question is at the core of the MCMC method and will be discussed next in the general setting of continuous state space.

11.1.2 Random Walks in $\mathbb{R}^n$

Since, in general, we are interested in problems in which the random variables take on values in $\mathbb{R}^n$ rather than in a finite state space as in the previous example, we are now going to set the stage for Markov chain Monte Carlo with non-discrete state space. We start by defining the concept of *random walk* in $\mathbb{R}^n$ which, as the name suggests, is a process of moving around by taking random steps. The most elementary random walk can be defined as follows:

1. Start at an arbitrary point $x^0 \in \mathbb{R}^n$.
2. Draw a random vector $w^1 \sim \mathcal{N}(0, \mathsf{I}_n)$ and set $x^1 = x^0 + \sigma w^1$.
3. Repeat the process: Set $x^{k+1} = x^k + \sigma w^{k+1}$, $w^{k+1} \sim \mathcal{N}(0, \mathsf{I}_n)$.

Using the random variables notation, the location of the random walk at time k is a realization of the random variable X^k, and we have an evolution model

$$X^{k+1} = X^k + \sigma W^{k+1}, \quad W^{k+1} \sim \mathcal{N}(0, \mathsf{I}_n).$$

The conditional density of X^{k+1}, given $X^k = x^k$, is

$$\pi_{X^{k+1}|X^k}(x^{k+1} \mid x^k) = \frac{1}{(2\pi\sigma^2)^{n/2}} \exp\left(-\frac{1}{2\sigma^2} \|x^k - x^{k+1}\|^2\right) = q_k(x^k, x^{k+1}).$$

The function $q_k : \mathbb{R}^n \times \mathbb{R}^n \to \mathbb{R}_+$ is called the *transition kernel* and it is the continuous equivalent of the transition matrix P in the discrete state space example. To establish this correspondence, consider the joint probability density of X^k and X^{k+1}, given by

$$\begin{aligned}
\pi_{X^k, X^{k+1}}(x^k, x^{k+1}) &= \pi_{X^{k+1}|X^k}(x^{k+1} \mid x^k)\pi_{X^k}(x^k) \\
&= q_k(x^k, x^{k+1})\pi_{X^k}(x^k).
\end{aligned}$$

By marginalizing the variable x^k, it follows that

$$\pi_{X^{k+1}}(x^{k+1}) = \int q_k(x^k, x^{k+1})\pi_{X^k}(x^k)dx^k,$$

that is, the transition kernel defines a linear mapping P that propagates the probability densities,

$$\mathsf{P} : \pi_{X^k} \mapsto \pi_{X^{k+1}},$$

through the integral equation above. Since

$$q_k(x, y) = q(x, y) = \frac{1}{(2\pi\sigma^2)^{n/2}}\exp\left(-\frac{1}{2\sigma^2}\|x - y\|^2\right) \quad \text{for all } k = 0, 1, 2, \ldots,$$

i.e., the step is always equally distributed independently of the value of k, the kernel q is called *time invariant*.

The process above defines a *chain* $\{X^k, \ k = 0, 1, \cdots\}$ of random variables, each with its own probability. The chain is a discrete time stochastic process with values in $\mathbb{R}^n$ and has the particular feature that the probability distribution of each variable X^{k+1} depends on the past only through the previous element X^k of the chain. This can be expressed in terms of conditional densities as

$$\pi_{X^{k+1}|X^0,\ldots,X^k}(x^{k+1} \mid x^0, x^1, \ldots, x^k) = \pi_{X^{k+1}|X^k}(x^{k+1} \mid x^k). \tag{11.6}$$

The condition (11.6) is the continuous version of the discrete condition (11.4), and it is referred to as the *Markov property*. As in the discrete case of the previous example, a stochastic process with this property is called a *Markov chain*.

Example 11.1 To understand the role of the transition kernel, consider a Markov chain defined by a random walk model in $\mathbb{R}^2$,

$$X^{k+1} = X^k + W^{k+1}, \quad W^{k+1} \sim \mathcal{N}(0, \mathsf{C}), \tag{11.7}$$

where $\mathsf{C} \in \mathbb{R}^{2\times 2}$ is a symmetric positive definite matrix, with eigenvalue decomposition

$$\mathsf{C} = \mathsf{U}\mathsf{D}\mathsf{U}^\mathsf{T}, \tag{11.8}$$

where $\mathsf{U} \in \mathbb{R}^{2\times 2}$ is an orthogonal matrix, and $\mathsf{D} \in \mathbb{R}^{2\times 2}$ is diagonal with positive diagonal entries. The inverse of C can be decomposed as

$$\mathsf{C}^{-1} = \mathsf{U}\mathsf{D}^{-1}\mathsf{U}^\mathsf{T} = \left(\mathsf{U}\mathsf{D}^{-1/2}\right)\underbrace{\left(\mathsf{D}^{-1/2}\mathsf{U}^\mathsf{T}\right)}_{=\mathsf{L}} = \mathsf{L}^\mathsf{T}\mathsf{L},$$

and the transition kernel can be written as

$$q(x^k \mid x^{k+1}) = \pi_{X^{k+1} \mid X^k}(x^{k+1} \mid x^k) \propto \exp\left(-\frac{1}{2}\|\mathsf{L}(x^k - x^{k+1})\|^2\right).$$

Alternatively, we may write the random walk model (11.7) as

$$X^{k+1} = X^k + \mathsf{L}^{-1}W^{k+1}, \quad W^{k+1} \sim \mathcal{N}(0, \mathsf{I}_2), \tag{11.9}$$

where the random step is whitened.

To illustrate the role of the covariance matrix, let

$$\mathsf{U} = \begin{bmatrix} \cos\theta & -\sin\theta \\ \sin\theta & \cos\theta \end{bmatrix}, \quad \theta = \frac{\pi}{6},$$

and

$$\mathsf{D} = \operatorname{diag}(s_1^2, s_2^2), \quad s_1 = 0.5, \ s_2 = 0.1.$$

In the random walk model (11.9), the standard deviation in the direction of the first eigenvector u_1 is five time the standard deviation in the orthogonal direction of u_2, hence we assume that the random steps have a component about five times larger in the direction of the first eigenvector u_1 than along the second eigenvector u_2, where

$$u_1 = \begin{bmatrix} \cos\theta \\ \sin\theta \end{bmatrix}, \quad u_2 = \begin{bmatrix} -\sin\theta \\ \cos\theta \end{bmatrix}.$$

The left panel of Fig. 11.3 shows a random walk realization with the covariance matrix $\mathsf{C} = \sigma^2 \mathsf{I}_2$ starting from the origin of $\mathbb{R}^2$ and choosing the step size $\sigma = 0.1$. In the right panel, the covariance matrix is chosen as above. Obviously, by judiciously choosing the transition kernel, we may guide the random walk quite effectively.

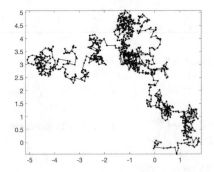 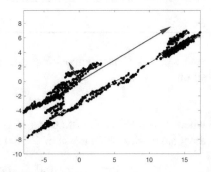

Fig. 11.3 Realizations of random walks. Left: The covariance matrix of the random step is $\sigma^2 \mathsf{I}_2$, with the standard deviation $\sigma = 0.1$, and the number of steps is $N = 1000$. Right: The covariance is C given in (11.8). The eigenvectors of the covariance matrix are indicated in the plot, scaled proportionally to the respective standard deviations

Consider now an arbitrary transition kernel $q : \mathbb{R}^n \times \mathbb{R}^n \to \mathbb{R}_+$, normalized so that

$$\int q(x, y)dy = 1.$$

Assume that X is a random variable with known probability density $\pi_X(x) = p(x)$. Suppose that we generate a new random variable Y by using the kernel $q(x, y)$, that is,

$$\pi_{Y|X}(y \mid x) = q(x, y).$$

The probability density of this new variable Y can be found by marginalization,

$$\pi_Y(y) = \int \pi_{Y|X}(y \mid x)\pi_X(x)dx = \int q(x, y)p(x)dx.$$

If the probability density of Y is equal to the probability density of X, i.e.,

$$\int q(x, y)p(x)dx = p(y), \tag{11.10}$$

we say that p is an *invariant density* of the transition kernel q. The classical problem in the theory of Markov chains can then be stated as follows: *Given a transition kernel q, find the corresponding invariant density p that satisfies Eq. (11.10).*

When using Markov chains to sample from a given density, we are actually considering the *inverse problem*: Given a probability density $p = p(x)$, generate a sample that is distributed according to it. If we had a transition kernel q with invariant density p, generating such sample would be easy: starting from x^0, draw x^1 from $q(x^0, x^1)$ considered as a probability density with respect to x^1 and x^0 fixed, and continue inductively. In general, given x^k, draw x^{k+1} from $q(x^k, x^{k+1})$. After a while, the x^k's generated in this manner are distributed more and more according to p. This was the strategy for generating the sample in the discrete state space problem of Subsect. 11.1.1.

Therefore the problem we are facing now is: *Given a probability density p, find a kernel q such that p is its invariant density.*

Probably the most popular technique for constructing such transition kernel is the Metropolis–Hastings algorithm.

11.2 Metropolis–Hastings Algorithm

We are now ready to derive, starting from the concept of invariant densities, one of the bread-and-butter Markov chain Monte Carlo (MCMC) algorithms, the Metropolis–Hastings (MH) algorithm.

11.2.1 Balance and Detailed Balance Equations

Assume that a probability density $p : \mathbb{R}^n \to \mathbb{R}_+$ is given and that the goal is to find a transition kernel such that p is its invariant probability density. To define a Markov chain, consider the following algorithm: Starting from the current point $x \in \mathbb{R}^n$, either

1. Stay put at x with probability $r(x)$, $0 \le r(x) < 1$, or
2. Move away from x using a kernel $K(x, y) \ge 0$, $x, y \in \mathbb{R}^n$.

The above alternatives can be summarized by defining a transition kernel consisting of a point mass and a distributed part,

$$q(x, y) = r(x)\delta_x(y) + K(x, y).$$

In order for q to define a transition kernel, it must satisfy

$$\int q(x, y)dy = r(x) \int \delta_x(y)dy + \int K(x, y)dy$$
$$= r(x) + \int K(x, y)dy = 1,$$

implying that

$$1 - r(x) = \int K(x, y)dy. \tag{11.11}$$

If the conditional density of the random variable Y, given $X = x$, is defined by using this transition kernel, we have, for any set $B \subset \mathbb{R}^n$,

$$P\{Y \in B \mid X = x\} = \int_B \pi_{Y|X}(y \mid x)dy$$
$$= \int_B q(x, y)dy$$
$$= r(x) \int_B \delta_x(y)dy + \int_B K(x, y)dy.$$

We observe that the first integral equals one if and only if $x \in B$, otherwise the integral vanishes. Therefore,

$$\int_B \pi_{Y|X}(y \mid x)dy = r(x)\chi_B(x) + \int_B K(x, y)dy, \tag{11.12}$$

where χ_B is the characteristic function (or indicator function) of the set B. Multiplying the left side of the identity (11.12) by $p(x)$, integrating over $\mathbb{R}^n$ and recalling that

$$\pi_Y(y) = \int \pi_{Y|X}(y \mid x)\pi_X(x)dx = \int \pi_{Y|X}(y \mid x)p(x)dx,$$

the left hand side yields

$$\int p(x) \int_B \pi_{Y|X}(y \mid x)dy = \int_B \left(\int \pi_{Y|X}(y \mid x)p(x)dx \right) dy$$

$$= \int_B \pi_Y(y)dy. \tag{11.13}$$

On the other hand, the right hand side gives

$$\int p(x) \left(r(x)\chi_B(x) + \int_B K(x, y)dy \right) dx \tag{11.14}$$

$$= \int_B \left(p(y)r(y) + \int p(x)K(x, y)dx \right) dy.$$

Since the identity of the right hand sides (11.13) and (11.14) holds for every B, we conclude that

$$\pi_Y(y) = p(y)r(y) + \int p(x)K(x, y)dx.$$

Since our goal is to find a transition scheme for which p is an invariant probability density, we substitute $\pi_Y(y) = p(y)$ in the expression above to get the necessary and sufficient condition

$$(1 - r(y))p(y) = \int p(x)K(x, y)dx.$$

Substituting the right-hand side of (11.11) with the roles of x and y interchanged in the left-hand side of the above identity, we get

$$p(y) \int K(y, x)dx = \int p(y)K(y, x)dx = \int p(x)K(x, y)dx. \tag{11.15}$$

This equation, known as the *balance equation*, is the necessary and sufficient condition for p to be the invariant density for the proposed transition scheme. This condition is satisfied, in particular, if the integrands are equal,

$$p(y)K(y, x) = p(x)K(x, y), \tag{11.16}$$

yielding a stronger, sufficient condition to guarantee the balance condition, known as the *detailed balance equation*. The Metropolis–Hastings algorithm provides a technique for finding a kernel K that satisfies (11.16).

11.2.2   Construction of the MH Transition

The Metropolis–Hastings algorithm starts by selecting a *proposal distribution*, or *candidate generating kernel* $R(x, y) \geq 0$, usually chosen so that generating a Markov chain with it is easy. It is mainly for this reason that Gaussian kernels are a very popular choice. Assume that R is a transition kernel, that is,

$$\int R(x, y)dy = 1.$$

If R satisfies the detailed balance equation,

$$p(y)R(y, x) = p(x)R(x, y),$$

we let $r(x) = 0$, hence $q(x, y) = K(x, y) = R(x, y)$, and we are done, since the previous analysis shows that p is an invariant density for this kernel. If, as is more likely to happen, the detailed balance equation is not satisfied, the left-hand side is either larger or smaller than the right-hand side. Assume, for the sake of definiteness, that

$$p(y)R(y, x) < p(x)R(x, y). \tag{11.17}$$

To enforce the detailed balance equation we modify the kernel K to

$$K(x, y) = \alpha(x, y)R(x, y),$$

where the correcting factor α is chosen so that

$$p(y)\alpha(y, x)R(y, x) = p(x)\alpha(x, y)R(x, y). \tag{11.18}$$

Since the kernel α need not be symmetric, we can choose

$$\alpha(y, x) = 1,$$

and let the other correcting factor be determined from (11.18):

$$\alpha(x, y) = \frac{p(y)R(y, x)}{p(x)R(x, y)} < 1. \tag{11.19}$$

Observe that if the direction of inequality (11.17) is reversed, we simply interchange the roles of x and y, letting instead $\alpha(x, y) = 1$. In summary, we define K as

$$K(x, y) = \alpha(x, y)R(x, y), \quad \alpha(x, y) = \min\left\{1, \frac{p(y)R(y, x)}{p(x)R(x, y)}\right\}.$$

This expression for K looks rather complicated, and it may seem that generating random draws should be all but simple. However, drawing from this kernel can be performed easily in two phases, as in the case of rejection sampling, according to the following algorithm.

Metropolis–Hastings algorithm: Given a proposal transition kernel $R(x, y)$, the probability density p, and the target size N of the sample:

1. **Initialize:** Choose x^0, set the counter $k = 1$.
2. **Iterate:** While $k < N$,

    (a) Given x^{k-1}, draw y using the transition kernel $R(x^{k-1}, y)$.
    (b) Calculate the *acceptance ratio*,

$$\alpha(x^{k-1}, y) = \frac{p(y)R(y, x^{k-1})}{p(x^{k-1})R(x^{k-1}, y)}.$$

    (c) Flip the α–coin: draw $t \sim \text{Uniform}([0, 1])$; if $\alpha > t$, accept y, and set $x^k = y$, otherwise stay put, setting $x^k = x^{k-1}$.
    (d) Advance the counter $k \to k + 1$.

Observe that rejection of the proposal leads to acceptance of the old point in the sample. Therefore, typically the sample contains multiple copies of some points. It is important to not discard the repetitions, as they reflect the importance of such points in representing the density. Alternatively, one can think that the repetitions are a way to give a larger importance weight to the sample point.

A careful reader will notice that the derivation given above is not a complete proof of the validity of the Metropolis–Hastings algorithm, because when the next point y was generated from the previous one, we tacitly assumed that the previous one, x, was sampled from the underlying p. We are omitting the proof that the generated distribution indeed converges toward p, because it requires more advanced technical tools that have not been introduced in this discussion. For a reference to the complete proof, see Notes and Comments.

Before presenting some computed examples, a few remarks are in order. Often, the proposal distribution R is chosen to be symmetric,

$$R(x, y) = R(y, x). \tag{11.20}$$

If this is the case, the acceptance ratio α simplifies to

$$\alpha(x, y) = \frac{p(y)R(y, x)}{p(x)R(x, y)} = \frac{p(y)}{p(x)}, \tag{11.21}$$

that is, α compares only the values of the density at the two points. Intuitively, the idea is that the move is always accepted if the proposed point is more probable than the old one, otherwise the move is accepted with a certain probability. This also explains the role of repeated sample points: a multiply repeated point is one with high probability, as it is hard to move away from it, and consequently, it deserves to have a larger weight.

11.2.3 Metropolis–Hastings in Action

We will now highlight some convenient features of the algorithm while following it into action, starting with an example.

Example 11.2 Consider the probability density in $\mathbb{R}^2$,

$$p(x) \propto \exp\left(-\frac{1}{2\sigma^2}((x_1^2 + x_2^2)^{1/2} - 1)^2 - \frac{1}{2\delta^2}(x_2 - 1)^2\right), \qquad (11.22)$$

where

$$\sigma = 0.1, \quad \delta = 1,$$

whose equiprobability curves are shown in Fig. 11.4.

We explore this density with a random walk sampler. For that purpose, consider the scaled white noise random walk proposal,

$$R(x, y) = \frac{1}{\sqrt{2\pi\gamma^2}}\exp\left(-\frac{1}{2\gamma^2}\|x - y\|^2\right).$$

Since the transition kernel is symmetric, satisfying condition (11.20), the acceptance ratio simplifies as in (11.21).

We start the Markov chain from the origin, $x^0 = (0, 0)$, and to illustrate how it progresses, we adopt the plotting convention that each new accepted point is

Fig. 11.4 The equiprobability curves of the original density (11.22)

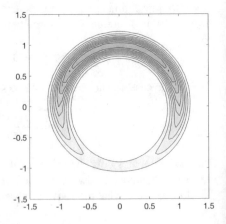

represented as a dot. If a proposal is rejected and we remain in the current position, the size of the dot increases, so that the area of the dot is proportional to the
number of rejections. A Matlab code that generates the sample can be written as
follows. For later use, we keep also track of how often a move is accepted.

```
nsample = 500;                % Size of the sample
Sample = zeros(2,nsample);
count = 1;
x = [0;0];                    % Initial point
Sample(:,1) = x;
lprop_old = logpdf(x);   % logpdf = log of the prob.density
acc_rate = 0;

while count < nsample
        % draw candidate
        y = x + step*randn(2,1);
        % check for acceptance
        lprop =  logpdf(y);
        if lprop - lprop_old > log(rand);
            % accept
            acc_rate = acc_rate + 1;
            x = y;
            lprop_old = lprop;
        end
        count = count+1;
        Sample(:,count+1) = x;
    end
```

We remark that in actual computations, the acceptance ratio is calculated in logarithmic form: we accept the move $x \to y$ if

$$\log p(y) - \log p(x) > \log t, \quad t \sim \text{Uniform}([0, 1]).$$

The reason for proceeding in this manner is to avoid numerical problems due to
underflow in the computation of the ratio of $p(y)$ and $p(x)$. In fact, it may happen
that the numbers $p(y)$ and $p(x)$ end up being smaller than the computing precision,
leading to an ill-defined ratio, while the logarithmic form avoids such problems.

In our first exploration of the density, we decide to move rather conservatively
and take very small steps by setting $\gamma = 0.02$. A plot of the first 500 points generated
by the MH algorithm is shown in Fig. 11.5. The figure also indicates how often the
proposal was accepted.

It is clear from the plot that, after 500 draws, the sampler has not even started to
explore the density. In fact, almost the entire sample has been used to move from the
initial point to the numerical support of the density. This initial tail, which has nothing
to do with the actual probability density, is usually referred to as the *burn-in* of the

Fig. 11.5 The
Metropolis–Hastings sample
with the step size $\gamma = 0.02$,
and 93% of the proposals are
accepted. The computed
sample mean is marked by
the cross hair

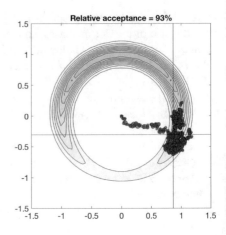

sample. It is normal procedure in MCMC sampling methods to discard the beginning of the sample to avoid that the burn-in affects the estimates that are subsequently calculated from the sample. In general, it is not easy to decide a priori how many points should be discarded.

The second observation is that the *acceptance rate* is rather high: approximately nine out of every ten proposed moves are accepted. A high acceptance rate usually indicates that the chain is moving too conservatively, and in that case longer steps should be used to get better coverage of the distribution.

Motivated by the observed high acceptance rate, we increase the step by a factor of hundred, choosing $\gamma = 2$. The results of this modification can be seen in Fig. 11.6. Now the acceptance rate is only 4%, meaning that most of the time the chain does not move and more than 90% of the proposals are rejected. Notice that the big dots in the figure indicate points from which the chain does not want to move away. In this case the burn-in effect is practically absent, and the estimated conditional mean is much closer to what one could expect. The result seems to suggest that low acceptance rate is better than too high.

By playing with the steplength, we may be able to *tune* the proposal distribution so as to achieve an acceptance rate between 20% and 30%, which is often advocated as optimal.

Figure 11.7 shows the results obtained with $\gamma = 0.5$, yielding an acceptance rate of approximately 24%. We see that the points are rather well distributed over the support of the probability density. The estimated mean, however, is not centered, probably because that the size of the sample is too small.

The previous example shows that the choice of the proposal distribution has an effect on the quality of the sample thus generated. While in the two-dimensional case it is fairly easy to assess the quality of the sampling strategy simply by looking at the scatter plot of the sample, in higher dimensions this approach becomes impossible, and more systematic means to analyze the sample are needed. There is the definitive

Fig. 11.6 The Metropolis–Hastings sample with the step size $\gamma = 2$. The computed sample mean is marked by the cross hair

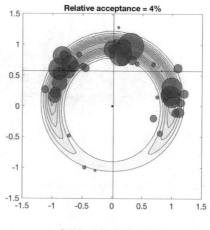

Fig. 11.7 The Metropolis–Hastings sample with the step size $\gamma = 0.5$

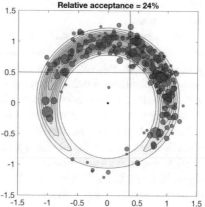

way to measure the quality of a sample, and we only touch on this rather complex topic here.

The Central Limit Theorem suggests a way to measure the quality of the sample. According to the Central Limit Theorem, the asymptotic convergence rate of a sum of N independently sampled, identically distributed random variables is $1/\sqrt{N}$. While the goal of Markov chain Monte Carlo sampling is to produce a sample that is asymptotically drawn from the limit distribution, focusing mostly on the *identically distributed* aspect, the *independence* of the computed sample is more problematic. Clearly, since the sample is a realization of a Markov chain, complete independence of the sample points cannot be expected: every draw depends on at least the previous element in the chain. This dependency has repercussions on the convergence of Monte Carlo integrals. Suppose that, on the average, only every kth sample point can be considered independent. Then, by the asymptotic law of the Central Limit Theorem, we may expect a convergence rate of the order of $\sqrt{k/N}$, a rate that is painfully slow if k is large. Therefore, when designing Metropolis–Hastings strategies, we should

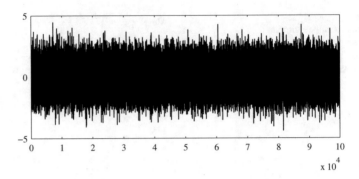

Fig. 11.8 A realization of a Gaussian white noise signal of length 100 000

aim at choosing the step length in the proposal distribution so that the *correlation length* is small.

Let us begin with a visual inspection of the samples. Suppose that we have a sampling problem in n spatial dimensions. While we cannot inspect a scatter plot when n is large, we may always look at the *sample histories* of individual components, plotting the individual components as a function of the sample index. The first question is then what a good sample history looks like.

A typical example of a sample with completely uncorrelated elements is a *white noise signal*: at each discrete time instant, the sample is drawn independently. A Gaussian white noise signal of length n is, in fact, a realization of a Gaussian multivariate vector with covariance I_n. Figure 11.8 shows a realization of a white noise signal. It looks like a "fuzzy worm." This gives a visual description to an MCMCer of what a good sample history should look like. This visual description, although rather vague, is in fact, quite useful to quickly assess the quality of a sample.

A more quantitative measure for assessing the quality of the sample can be obtained by looking at its correlation structure. The *autocorrelation* of a signal is a useful tool to analyze the independency of realizations. Let z_j, $1 \leq j \leq N$ denote a discrete time finite segment of a real-valued signal. Assume, for simplicity, that the signal has zero mean. After augmenting the signal with trailing zeros to an infinite signal, $1 \leq j < \infty$, consider the discrete convolution,

$$h_k = \sum_{j=1}^{\infty} z_{j+k} z_j, \quad k = 0, 1, \ldots$$

When $k = 0$, the above formula returns the total energy of the signal,

$$h_0 = \sum_{j=1}^{N} z_j^2 = \|z\|^2.$$

If z is a white noise signal and $k > 0$, the random positive and negative contributions cancel out, and $h_k \approx 0$. This observation gives a natural tool to analyze the independency of the components in a sample: plot

$$k \mapsto \widehat{h}_k = \frac{1}{\|z\|^2} h_k, \quad k = 0, 1, \ldots$$

and estimate the correlation length from the rate of decay of this sequence. The quantity $\widehat{h}_k$ is the autocorrelation of z with *lag k*.

The following example demonstrates how to use this idea.

Example 11.3 Consider the horseshoe distribution of the previous example, and the Metropolis–Hastings algorithm with a scaled white noise proposal distribution. We consider three different step sizes: $\gamma = 0.1$, $\gamma = 1$, and $\gamma = 5$. We generate a sample $\{x^1, x^2, \ldots, x^N\}$ of size $N = 50\,000$, where, for a good measure, we have discarded the 500 first sample points as they may not represent the distribution. We calculate the mean,

$$\overline{x} = \frac{1}{N} \sum_{j=1}^{N} x^j,$$

and the lagged autocorrelations of the centered components,

$$\widehat{h}_{i,k} = \frac{1}{\|z\|^2} \sum_{j=1}^{N-k} z_{j+k} z_j, \quad z_j = (x^j - \overline{x})_i, \quad i = 1, 2.$$

Figure 11.9 shows the sample history of both components x_1^j and x_2^j with the different step sizes. Visually, the most reasonable step size $\gamma = 1$ produces a sample history that is similar to the white noise sample, while with the smaller step size there is a visible low frequency component that tells us about the slow walk around the density. The larger step size, on the other hand, causes the sampler to stay put for long periods of times with an adverse effect on the independency of the samples. The computed autocorrelation functions agree with the visual assessment, the step size $\gamma = 1$ producing the most rapidly decreasing autocorrelation function. Based on the autocorrelation function, one can conclude that roughly every 50th sample point can be safely considered to represent an independent sample from the distribution.

We close this section with a computed example of a parameter estimation problem and its solution using the Metropolis–Hastings algorithm.

Example 11.4 Consider a system of chemical reactions involving substances A, B, C, and D,

$$A \underset{k_2}{\overset{k_1}{\rightleftharpoons}} B \underset{k_4}{\overset{k_3}{\rightleftharpoons}} C + D,$$

where k_j, $1 \le j \le 4$ are the reaction rates. Assuming that the reactions satisfy the mass action model, the concentrations $u_1 = [A]$, $u_2 = [B]$, $u_3 = [C]$ and $u_4 = [D]$

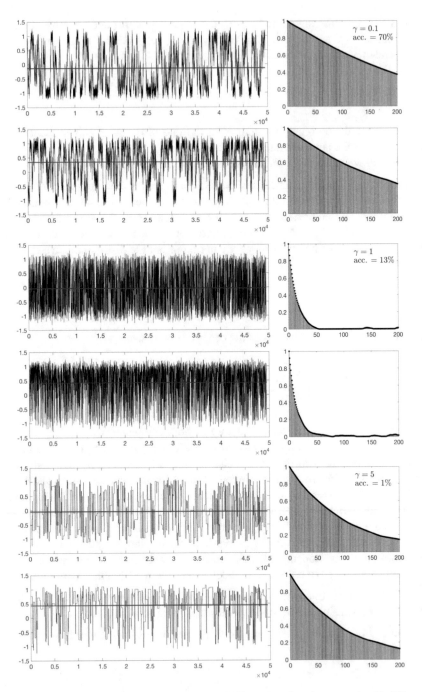

Fig. 11.9 Sample histories of the components x_1 (odd rows) and x_2 (even rows) with different proposal step parameters γ, and the corresponding autocorrelation functions. The red horizontal lines indicate the estimated posterior means

of the substances satisfy the system of differential equations

$$\frac{du_1}{dt} = -k_1 u_1 + k_2 u_2,$$

$$\frac{du_2}{dt} = k_1 u_1 - k_2 u_2 - k_3 u_2 + k_4 u_3 u_4,$$

$$\frac{du_3}{dt} = k_3 u_2 - k_4 u_3 u_4,$$

$$\frac{du_3}{dt} = k_3 u_2 - k_4 u_3 u_4.$$

The problem considered here is to estimate the reaction rates k_j from noisy observations of the concentrations, given that the initial concentrations are known. Denoting by $k \in \mathbb{R}^4$ the vector containing the reaction rates and by $u(t) \in \mathbb{R}^4$ the vector of the concentrations at time t, the data are defined as

$$b_j = u(t_j) + e_j, \quad 1 \leq j \leq m, \quad u(0) = u_0 \text{ given},$$

where $0 < t_1 < \ldots < t_m$. For simplicity, we assume that e_j is a realization of a random variable $E_j \sim \mathcal{N}(0, \sigma^2 I_4)$, and the variables E_j are mutually independent.

A priori we only assume that the reaction rates are positive. To write a proper prior, however, we need to impose an upper bound constraint, $k_j < \kappa_j$ for some large value κ_j, leading to a prior model

$$\pi_K(k) \propto \prod_{j=1}^{4} \chi_{[0,\kappa_j]}(k_j),$$

which is the product of the characteristic functions of the respective intervals.

We generate the data with the above generative model, using parameter values

$$k_1^* = 0.2, \quad k_2^* = 0.5, \quad k_3^* = 5, \quad k_4^* = 15,$$

and initial values

$$u_{0,1} = 0.2, \quad u_{0,2} = 1, \quad u_{0,3} = 0.2, \quad u_{0,4} = 0.05.$$

The data are collected at times

$$t_j = 0.01 + (j - 1)0.03, \quad 1 \leq j \leq 17,$$

and to each computed concentration, normally distributed error with standard deviation $\sigma = 0.01$ is added. The data are shown in Fig. 11.10.

The likelihood model in this case is given by

Fig. 11.10 The solid curves correspond to the noiseless concentrations, and the data are obtained by adding normal error of standard deviation $\sigma = 0.01$ to the sampled values at times t_j

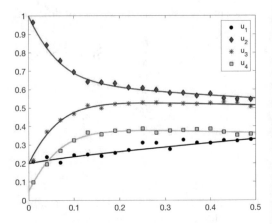

$$\pi_{B|K}(b \mid k) \propto \exp\left(-\frac{1}{2\sigma^2} \sum_{j=1}^{17} \|b_j - u(t_j; k)\|^2\right);$$

for practical computations, $u(t_j)$ can be computed with any standard ODE solver such as `ode45` in Matlab.

To estimate the parameters, we run the standard Metropolis–Hastings algorithm with Gaussian random walk proposal, step size $\gamma = 0.1$, starting the chain at $k_0 = [1; 1; 10; 10]$, and we generate a chain of length $N = 200\,000$. The acceptance rate with this step size turns out to be 5.8%. Reducing the step size to $\gamma = 0.05$ leads to an acceptance rate of 17.2%. Based on visual inspection of the time traces, we discard 500 sample points from the beginning of the chain.

In Fig. 11.11 the results are presented graphically as a *scatter matrix*, showing the histograms of each component in the diagonal panels, and pairwise scatter plots in the off-diagonal panels.

The scatter plots of the components k_1 versus k_2, and k_3 versus k_4, reveal a strong correlation between them, indicating that the data contain information about the ratios of the corresponding rate constants. We point out that the ratios k_1/k_2 and k_3/k_4 are related to the equilibrium conditions that the system asymptotically approaches, indicating that the concentrations at large times convey information about those ratios. In our example, the system was still relatively far from the equilibrium.

11.3 Gibbs Sampler

Another classical MCMC algorithm is Gibbs sampler. The intuitive idea behind the algorithm is rather simple, and it carries a certain similarity with the classical

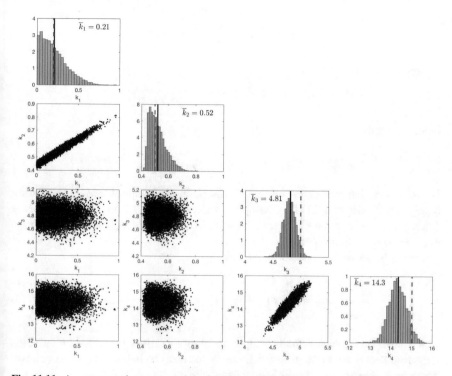

Fig. 11.11 A scatter matrix representation of the Metropolis–Hastings run with proposal step size $\gamma = 0.1$. In the panels along the diagonal, the histograms of each component is shown, with the sample mean marked by a solid black line. For comparison, the generative value is marked by a dashed red line. The off-diagonal panels show the pairwise scatter plots of the components. The results with the proposal steps size $\gamma = 0.05$ are essentially similar, the sample means coinciding up to the third decimal

Gauss–Seidel algorithm for solving linear systems: The random draws are done component-wise, updating one coordinate at the time.

Before presenting the details of the algorithm, we establish the notation. Given a probability density in $\mathbb{R}^n$, $p(x) = p(x_1, x_2, \ldots, x_n)$, we use the expression

$$p(x_j \mid x_1, \ldots, x_{j-1}, x_{j+1}, \ldots, x_n)$$

to denote the density of the component x_j conditioned on the other components, obtained by

1. Freezing all components x_ℓ, except for $\ell = j$;
2. Normalizing the resulting density to have integral with respect to x_j equal to one.

More precisely, denoting by $p(x_j)$ the marginal density,

$$p(x_j) = \int_{\mathbb{R}^{n-1}} p(x_1, \ldots, x_n) dx_1 \cdots dx_{j-1} dx_{j+1} \cdots dx_n,$$

we define

$$p(x_j \mid x_1, \ldots, x_{j-1}, x_{j+1}, \ldots, x_n) = \frac{p(x_1, \ldots, x_n)}{p(x_j)},$$

or, equivalently,

$$p(x) = p(x_j \mid x_1, \ldots, x_{j-1}, x_{j+1}, \ldots, x_n) p(x_j).$$

Unlike in the Metropolis–Hastings algorithm where the sampling was done by using a proposal distribution, in Gibbs sampling we draw directly from the target density $p(x)$, $x \in \mathbb{R}^n$. Consequently, there is no need to define an acceptance probability, as every proposal will be accepted.

Formally, we define a transition kernel

$$q(x, y) = \prod_{i=1}^{n} p(y_i \mid y_1, \ldots, y_{i-1}, x_{i+1}, \ldots, x_m), \tag{11.23}$$

which leads to the conceptually simple updating algorithm presented below. This transition kernel does not, in general, satisfy the detailed balance equation (11.16), however, it satisfies the weaker balance equation,

$$\int_{\mathbb{R}^n} p(y) q(y, x) dx = \int_{\mathbb{R}^n} p(x) q(x, y) dx, \tag{11.24}$$

which has been shown to be the necessary and sufficient condition for p to be an invariant distribution of the transition rule defined by the kernel q. The proof is straightforward but tedious. In order to avoid that the discussion becomes overly technical, we present it for the case $n = 2$, with the understanding that similar ideas apply to the general case.

For $n = 2$, the transition kernel is given by

$$q(x, y) = p(y_1 \mid x_2) p(y_2 \mid y_1),$$

and therefore

$$q(y, x) = p(x_1 \mid y_2) p(x_2 \mid x_1).$$

We start by considering the left-hand side of the identity (11.24). Integrating $q(y, x)$ with respect to x yields

$$\int_{\mathbb{R}^2} q(y,x)dx = \int_{\mathbb{R}^2} p(x_1 \mid y_2)p(x_2 \mid x_1)dx_1 dx_2$$

$$= \int_{\mathbb{R}} \left(p(x_1 \mid y_2) \underbrace{\int_{\mathbb{R}} p(x_2 \mid x_1)dx_2}_{=1} \right)dx_1$$

$$= \int_{\mathbb{R}} p(x_1 \mid y_2)dx_1 = 1,$$

hence

$$\int_{\mathbb{R}^2} p(y)q(y,x)dx = p(y).$$

Next write

$$p(x)q(x,y) = p(x)p(y_1 \mid x_2)p(y_2 \mid y_1) = p(x_1,x_2)p(y_1 \mid x_2)p(y_2 \mid y_1),$$

and integrate with respect to x_1 to obtain

$$\int_{\mathbb{R}} p(x)q(x,y)dx_1 = p(y_1 \mid x_2)p(y_2 \mid y_1)\underbrace{\int_{\mathbb{R}} p(x_1,x_2)dx_1}_{=p(x_2)}$$

$$= \underbrace{p(y_1 \mid x_2)p(x_2)}_{=p(y_1,x_2)} p(y_2 \mid y_1)$$

$$= p(y_1,x_2)p(y_2 \mid y_1).$$

Integrating this expression with respect to x_2, we obtain

$$\int_{\mathbb{R}} p(y_1,x_2)p(y_2 \mid y_1)dx_2 = p(y_2 \mid y_1)\underbrace{\int_{\mathbb{R}} p(y_1,x_2)dx_2}_{p(y_1)}$$

$$= p(y_2 \mid y_1)p(y_1)$$

$$= p(y_1,y_2) = p(y),$$

thus completing the proof.

As intimidating as formula (11.23) may look, in practice it is quite easy to implement it, by treating the factors in the product sequentially, until every component has been updated, see Fig. 11.12. This is summarized in the following sampling algorithm.

Gibbs Sampler algorithm: Given the probability density p and the target size N of the sample of its realizations:

1. **Initialize:** Choose x^0, set the counter $k = 1$.
2. **Iterate:** While $k < N$,

Fig. 11.12 Gibbs Sampler
generates the sample by
drawing the coordinates one
at a time from the marginal
density, keeping the
remaining coordinates fixed

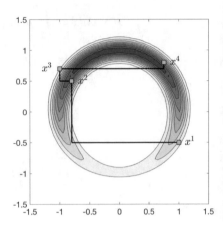

(1) Draw x_1^k from $t \mapsto p(t, x_2^{k-1}, x_3^{k-1}, \cdots, x_n^{k-1})$,

(2) Draw x_2^k from $t \mapsto p(x_1^k, t, x_3^{(k-1)}, \cdots, x_n^{(k-1)})$,

$\vdots$

(n) Draw x_n^k from $t \mapsto p(x_1^k x_2^k, \cdots, x_{n-1}^k, t)$.
     Increase $k \to k+1$.

A major appeal of the Gibbs sampler algorithm is that there is no need for tuning
parameters affecting the acceptance as in the Metropolis–Hastings algorithm. The
drawback, however, is that the drawing from one-dimensional marginals, while conceptually straightforward, may be tedious to implement and time consuming. This is
particularly true if the marginals are multi-modal, or when the support of the density
is poorly known. Moreover, it is easy to imagine situations in which Gibbs sampler
fails to explore the density, or when the process becomes very slow. An example of
the former situation is given by a bimodal distribution in $\mathbb{R}^2$,

$$p(x) \propto \chi_{B_1}(x) + \chi_{B_2}(x),$$

where $B_1, B_2 \subset \mathbb{R}^2$ are two sets in the plane so that

$$B_1 \subset \{(x_1, x_2) \mid x_1, x_2 < 0\}, \quad B_2 \subset \{(x_1, x_2) \mid x_1, x_2 > 0\}.$$

see the left panel of Fig. 11.13. In this case, it is impossible to reach B_2 starting from
B_1 and vice versa by the coordinate update scheme. An example of the latter case
is given by the distribution shown in the right panel of the same figure, where the
updating step size in the coordinate directions is very small, making the progress
of the exploration prohibitively slow. Here a simple and educated rotation of the
coordinate axes would resolve the problem, however, it may be hard to find, in
particular when n is large. Another option is to perform the updates in random

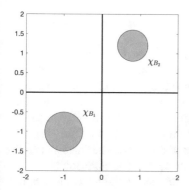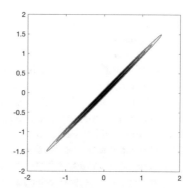

Fig. 11.13 Two examples of probability densities that Gibbs sampler has problems to explore. On the left, the sampler cannot jump from one disc to the other along paths with segments parallel to coordinate axes. On the right, the components x_1 and x_2 are strongly correlated, and the distribution allows only short steps parallel to the coordinate axes, therefore rendering the sampler inefficient

directions rather than along coordinate directions. The ensuing algorithm is referred to as *Hit and Run algorithm*.

11.4 Preconditioned Crank–Nicholson

In the Metropolis–Hastings algorithm, the acceptance rate can be controlled by the step size in the proposal distribution, with smaller steps leading to higher acceptance rate. The tuning process is known to be sensitive to the dimensionality of the underlying space and, in high dimensions, targeting a given acceptance rate leads typically to a step size so small that the algorithm becomes impractical, requiring sample sizes that are not realistically achievable. A solution proposed in the literature for a particular class of problems is to modify the proposal, and its tuning parameter, leading to a sampling scheme referred to as *preconditioned Crank–Nicholson (pCN)* algorithm.

Following the original presentation of the pCN algorithm, we assume that the probability density p to be explored is of the form

$$p(x) \propto e^{-\Phi(x)} p_0(x), \tag{11.25}$$

where p_0 is a Gaussian probability distribution with zero mean and symmetric positive definite covariance $C \in \mathbb{R}^{n \times n}$,

$$p_0(x) = \mathcal{N}(x \mid 0, C),$$

and Φ is a real-valued potential function. In this case, we say that the Gaussian p_0 is a dominant distribution. Observe that the formulation is particularly natural for posterior distributions with a Gaussian prior p_0, the exponential representing in this case the likelihood density.

We start by observing that since p_0 is Gaussian, drawing samples from p_0 is straightforward by using the Mahalanobis transformation. Therefore, one could use rejection sampling with p_0 as a proposal. This, however, is usually not a good idea, since rejection sampling may lead to extremely low acceptance rates, in particular, when the dimensionality of the problem is high, or when the data are of good quality. An advantage of MCMC algorithms is that, unlike rejection sampling, they learn the effective support of the density, and it is possible to control the step size so that the proposals do not fall too far from of the previously accepted sample points.

A natural way to take advantage of this idea is to modify the rejection sampling by introducing a control on the step size and to consider a Metropolis–Hastings algorithm with the dominant Gaussian distribution as the proposal density. Define the random walk proposal

$$y = x^{k-1} + \beta w^k, \quad w^k \sim \mathcal{N}(0, \mathbf{C}) \tag{11.26}$$

with some $\beta > 0$ controlling the step size. To get more insight into the problem, assume first that $\Phi(x) = 0$ and $p(x) = p_0(x)$, that is, we are simply seeking to generate a sample from the density p_0 itself. Assuming that $X \sim p_0$, consider a random variable corresponding to the proposal (11.26),

$$Y = X + \beta W, \quad X, W \sim \mathcal{N}(0, \mathbf{C}), \tag{11.27}$$

assuming that X and W are mutually independent. Then, Y is a zero mean Gaussian random variable, with covariance

$$\mathrm{E}(YY^\mathsf{T}) = \mathrm{E}(XX^\mathsf{T}) + \beta^2 \mathrm{E}(WW^\mathsf{T}) = (1 + \beta^2)\mathbf{C},$$

or, equivalently,

$$Y \sim \mathcal{N}(0, (1 + \beta^2)\mathbf{C}).$$

Therefore, the proposal (11.26) is drawn from the dominant Gaussian distribution with covariance inflated by a factor $(1 + \beta^2)$, and the acceptance of the proposal is not automatic. In fact, when the dimensionality of the problem increases, the acceptance rate drops quickly, and one needs to decrease the step size β, which makes the algorithm inefficient. However, we can use the same remedy for the covariance inflation that we used in Sect. 4.7.2 when discussing the weighted bootstrap algorithm. More precisely, we modify formula (11.27) to

$$Y = \sqrt{1 - \beta^2} X + \beta W,$$

by moving the point X slightly closer to the mean of p_0. Now, the covariance of Y is

$$E(YY^{\mathsf{T}}) = (1 - \beta^2)E(XX^{\mathsf{T}}) + \beta^2 E(WW^{\mathsf{T}}) = \mathsf{C},$$

that is, the proposal is drawn from p_0 and is automatically accepted!

After these preliminary considerations, consider the Metropolis–Hastings acceptance ratio corresponding to the proposal (11.26). The move to y is accepted, and $x^k = y$, if

$$\frac{p(y)}{p(x^{k-1})} > t, \quad t \sim \text{Uniform}([0, 1]),$$

otherwise we set $x^k = x^{k-1}$. Taking into account the particular form of the probability density, the acceptance condition can be written in logarithmic form as

$$\frac{1}{2}\|x^{k-1}\|_{\mathsf{C}}^2 + \Phi(x^{k-1}) - \left(\frac{1}{2}\|y\|_{\mathsf{C}}^2 + \Phi(y)\right) > \log t.$$

Here, the notation $z^{\mathsf{T}}\mathsf{C}^{-1}z = \|z\|_{\mathsf{C}}^2$ is used, which differs from the similar notation used in Chap. 9. Another way of expressing the proposal is to write the conditional density of the random variable Y given the realization of X^{k-1},

$$\pi_{Y|X^{k-1}}\left(y \mid x^{k-1}\right) \sim \mathcal{N}(x^{k-1}, \beta^2\mathsf{C}).$$

Modification of the proposal distribution, with a given β, $0 < \beta < 1$, yields

$$\pi_{Y|X^{k-1}}\left(y \mid x^{k-1}\right) \sim \mathcal{N}(\sqrt{1 - \beta^2}x^{k-1}, \beta^2\mathsf{C}),$$

and write the proposal as

$$y = \sqrt{1 - \beta^2}x^{k-1} + \beta w^k, \quad w^k \sim \mathcal{N}(0, \mathsf{C}).$$

Observe that since this proposal kernel is not symmetric, the acceptance ratio needs to be calculated using the formula (11.19). Writing first the transition kernel as

$$\pi_{Y|X^{k-1}}\left(y \mid x^{k-1}\right) = R(x, y) \propto \exp\left(-\frac{1}{2\beta^2}\|y - \sqrt{1 - \beta^2}x\|_{\mathsf{C}}^2\right),$$

we see that the ratio of the proposal kernels is

$$\frac{R(y, x)}{R(x, y)} = \exp\left(-\frac{1}{2\beta^2}\|x - \sqrt{1-\beta^2}y\|_C^2 + \frac{1}{2\beta^2}\|y - \sqrt{1-\beta^2}x\|_C^2\right)$$

$$= \exp\left(-\frac{1}{2\beta^2}\left(\|x\|_C^2 - 2\sqrt{1-\beta^2}x^TC^{-1}y + (1-\beta^2)\|y\|_C^2\right.\right.$$

$$\left.\left. - \|y\|_C^2 + 2\sqrt{1-\beta^2}y^TC^{-1}y - (1-\beta^2)\|x\|_C^2\right)\right)$$

$$= \exp\left(-\frac{1}{2}\left(\|x\|_C^2 - \|y\|_C^2\right)\right).$$

Likewise, the ratio of the densities is

$$\frac{p(y)}{p(x)} = \exp\left(-\frac{1}{2}\|y\|_C^2 - \Phi(y) + \frac{1}{2}\|x\|_C^2 + \Phi(x)\right),$$

so that the acceptance ratio is reduced to

$$\alpha(x, y) = \exp\left(-\Phi(y) + \Phi(x)\right),$$

which depends solely on the likelihood! The fact that the Gaussian portion is not playing a role in the acceptance ratio is a major reason for the popularity of the pCN algorithm for high-dimensional problems.

We summarize the pCN steps in the following algorithm.

Preconditioned Crank–Nicholson algorithm: Given a probability density p of the form (11.25), and the target size N of the sample:

1. **Initialize:** Choose x^0, set the counter $k = 1$.
2. **Iterate:** While $k < N$,

   (a) Given x^{k-1}, define

   $$y = \sqrt{1-\beta^2}x^{k-1} + \beta w^k, \quad w \sim \mathcal{N}(0, C).$$

   (b) Calculate the acceptance ratio,

   $$\alpha(x^{k-1}, y) = \exp\left(-\Phi(y) + \Phi(x^{k-1})\right).$$

   (c) Flip the α–coin: draw $t \sim$ Uniform$([0, 1])$; if $\alpha > t$, accept y, and set $x^k = y$, otherwise stay put, setting $x^k = x^{k-1}$.
   (d) Advance the counter $k \rightarrow k + 1$.

Notice that the parameter β controls the acceptance rate: In the limit, as $\beta \rightarrow 0+$, the proposal y converges to the previous sample vector, and therefore the acceptance ratio converges to one. As a rule of thumb, the acceptance rates advocated in the literature for pCN should be below or around 50%.

Notes and Comments

For a reference to Frobenius–Perron Theorem and related topics in linear algebra, we refer to the book [56]. There is a vast literature on Markov chain Monte Carlo methods, addressing both theoretical and practical aspects. For good collections of articles addressing several aspects of MCMC, see [6, 38]. The origins of the Metropolis–Hastings algorithm go back to the works of Nikolas Metropolis and colleagues on simulated annealing, [55], the current version being introduced by Hastings in [43]. For the origins of Gibbs sampler, we refer to [33, 34]. In the discussion of the MCMC algorithm, we have left out a number of important details needed to guarantee that the algorithm indeed explores the underlying distribution. For a concise but comprehensive discussion, we refer to [59].

The basic Metropolis–Hastings algorithm can be made more efficient by using proposal distributions with size and spatial orientation adjusted to the underlying density that is explored. *Adapted Metropolis* (AM) algorithm [39] is dynamically updating the Gaussian proposal distribution by recalculating the covariance matrix empirically from the previously accepted sample points, leading usually to higher acceptance of the proposals and therefore improved computational efficiency.

A different idea to improve the proposal distribution is to guide the proposals toward the local maxima of the underlying density. *Metropolis-adjusted Langevin algorithm* (MALA) [64] uses a discretized Langevin diffusion process as a proposal, where the random step drawn from the Gaussian distribution is biased toward the direction of the gradient of the negative logarithm of the density. The proposal therefore is closely related to the stochastic gradient descent algorithm for finding a minimum of a differentiable function. The Langevin algorithm is related to *Hybrid Monte Carlo* (HMC) algorithm having its roots in molecular dynamics simulations [30]. The proposal is a discrete approximation of Hamiltonian dynamics [58], for which HMC is also an abbreviation for *Hamiltonian Monte Carlo*. As comprehensive references to various MCMC algorithms we refer to [6, 52].

For a detailed introduction to the Preconditioned Crank–Nicholson algorithm, we refer to the articles [25, 40].

Chapter 12
Dynamic Methods and Learning from the Past

Richard Price, a Welsh philosopher and mathematician who has significantly contributed to the early development of Bayesian theory, published in 1764 an essay on Bayes' work, in which he asked how to assign a subjective probability to the sunrise, given that the sun had been observed to rise a given number of times before. Price's idea is that we learn from earlier experiences, and update our expectations based on them. The question was revisited by Pierre-Simon Laplace in his 1774 essay, and again in 1777 by the French scientist and mathematician George-Louis Leclerc de Buffon. While this interest in the question of sunrise may seem rather academic, its real value is the scientist's probing of the human thought process, and the difficulty of translating qualitative beliefs into numbers, a major challenge of Bayesian science even today. This quest also serves as an introduction to sequential Bayesian methods: *Today's posterior will be tomorrow's prior while waiting for the new observation to arrive.*

12.1 The Dog and the Hunter

We start with an introductory example to clarify the concept of Bayesian filtering and particle methods in general.[1]

Consider the following problem: A hunter, in full mimetic gear, leaves the house with his loyal dog to take a walk in the forest. We want to reconstruct the trajectory of the hunter, who cannot be observed, based on occasional sightings of the dog. What we know is that

1. At $t = 0$, the hunter left from his house.
2. At times $0 < t_1 < t_2 < \ldots < t_n$, the dog was seen at positions $q_1, q_2, \ldots, q_n$.

---

[1] We owe this example to our colleague Giulio d'Agostini.

© The Author(s), under exclusive license to Springer Nature Switzerland AG 2023
D. Calvetti and E. Somersalo, *Bayesian Scientific Computing*, Applied Mathematical
Sciences 215, https://doi.org/10.1007/978-3-031-23824-6_12

3. The speed of the hunter is likely to be no greater than a reasonable walking speed $v_{max} > 0$.
4. The dog usually does not venture further than a distance $d_{max} > 0$ from its master.

We solve the problem by generating, at each time step, a large cloud of possible positions of the hunter, propagate them according to the information we have, and, at each sighting of the dog, assess the viability of each particle.

We start by first treating the problem heuristically, and afterwards we proceed to put our solution on more solid ground. Let N denote the number of *particles*, which are realizations of the $\mathbb{R}^2$-valued random variable X_t modeling the position of the hunter at the time instance t. Since at $t = 0$ we know for sure that the hunter is at home, positioned at the coordinate origin of $\mathbb{R}^2$, we initialize the problem by choosing N particles

$$x_0^1 = x_0^2 = \ldots = x_0^N = 0,$$

all drawn from the prior probability density

$$\pi_{X_0}(x) = \delta_0(x).$$

Since the particles are identical, we assign them equal weights,

$$w_0^1 = w_0^2 = \ldots = w_0^N = \frac{1}{N}.$$

Next, we start to propagate. Denoting by x_1^ℓ the position of the ℓth particle at time $t = t_1$, we set

$$\widehat{x}_1^\ell = x_0^\ell + v_1^\ell, \quad 1 \leq \ell \leq N, \tag{12.1}$$

where v_1^ℓ is a random step, reflecting the fact that we do not know which way the hunter moves. If we interpret v_{max} as the maximum speed of the hunter, we should have

$$\|v_1^\ell\| \leq v_{max}(t_1 - t_0) = \gamma_1.$$

If there is no information on preferred direction, the direction of the move is arbitrary, and we may define v_1^ℓ as a realization of a random variable V_1 defined as

$$V_1 = \gamma_1 S \begin{bmatrix} \cos \Theta \\ \sin \Theta \end{bmatrix}, \quad S \sim \text{Uniform}([0, 1]), \quad \Theta \sim \text{Uniform}([0, 2\pi]).$$

This way, the particles (12.1) define a predictive sample, $\{\widehat{x}_1^1, \ldots, \widehat{x}_1^N\}$, which is based solely on the model, without using any data. The predictive sample can be thought of as our a priori information of the position of the hunter at $t = t_1$.

At time $t = t_1$, the first observation of the dog's position arrives. Therefore, we need to define a likelihood expressing the distribution of the position of the dog, assuming that we know where the hunter is. If the only information that we have is that the dog is never more than d_{max} away from the master, a natural candidate for the likelihood is

$$\pi_{Q_1|X_1}(q_1 \mid x_1) = \frac{1}{\pi d_{max}^2} \chi_{B(x_1,d_{max})}(q_1) = \begin{cases} 1/\pi d_{max}^2, & \text{if } \|d_1 - x_1\|\} < d_{max}, \\ 0 & \text{otherwise}, \end{cases}$$

(12.2)

the characteristic function of the disk centered at x_1 with radius d_{max}. Heuristically, it looks reasonable to associate to each particle x_1^ℓ the weight w_1^ℓ, with

$$w_1^\ell = \pi_{Q_1|X_1}(q_1 \mid x_1^\ell), \quad q_1 = \text{the observed value of } q_1,$$

(12.3)

and then normalize the weights so that they sum to one,

$$w_1^\ell \to \frac{w_1^\ell}{\sum_{\ell'} w_1^{\ell'}}.$$

With the likelihood model (12.2), the weights (12.3) before normalization are either $1/\pi d_{max}^2$ if the point is inside the disk, and zero otherwise.

Next we perform importance sampling: Select N indices $\ell_1, \ell_2, \cdots, \ell_N$, drawn with replacement from the set $\{1, 2, \ldots, N\}$, where each integer ℓ has probability w_1^ℓ of being selected, and define the new sample as

$$x_1^k = \widehat{x}_1^{\ell_k}, \quad 1 \le k \le N.$$

Observe that this way, the particles with zero weight that we deemed to be impossible positions for the hunter are never chosen, while positions more likely to explain the data may be selected multiple times. The weights of the new particles are all equal, $w_2^\ell = 1/N$, because importance is now expressed in the form of representation, that is, more probable positions may have been picked multiple times. The steps are explained in a graphical form in Fig. 12.1. The process can now be repeated for the next time step.

Before discussing the details of the proposed algorithm, let us consider a small computed example illustrating the merits and flaws of the approach.

Example 12.1 In our example, we generate the data assuming that the velocity of the hunter is given by

$$v(t) = \cos 3\pi t \begin{bmatrix} \cos \pi t \\ \sin \pi t \end{bmatrix}, \quad 0 \le t \le 1,$$

satisfying

$$\|v(t)\| \le 1 = v_{max}.$$

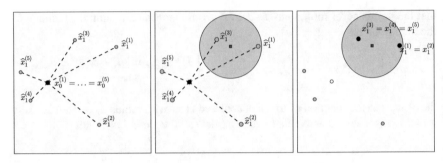

Fig. 12.1 Left: The initial sample consists of five particles indicating the hunter's initial position. Each particle is propagated by adding a random step, resulting in proposal particles $\widehat{x}_1^{(j)}$. Middle: The dog shows up (red square). The blue disk indicates all possible locations no more than $d_{\max}$ away from the dog. Only two proposal particles qualify. Right: New sample of hunter's position is drawn. Since the proposal particles outside the disk have zero weight, all five points must be replicates of the two possible positions

The trajectory of the hunter is shown in Fig. 12.2. We assume that at times $t_j = j/30$, $1 \leq j \leq 30$, the dog is observed at distance at most $d_{\max} = 6$ from the hunter. The red dots in the figure indicate the observed positions of the dog, joined with a line segment to the corresponding position of the hunter.

To highlight the limitations of the proposed model, we first run the algorithm with a small number of particles, setting $N = 50$. In Fig. 12.3, a few snapshots of the progression are shown: In the figure, the predictive particles $\widehat{x}_k^\ell$ are plotted with the color coding indicating if the weight w_k^ℓ is zero or positive. The snapshots clearly demonstrate the impoverishment of the sample over time; sometimes, only one of the particles has a non-vanishing weight, and at the final step, none of the particles survive, leading to a failure of the algorithm.

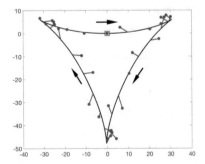

Fig. 12.2 Generative trajectory of the hunter with the positions at which the dog (right) is observed at times $t_j = j/30$, $1 \leq j \leq 30$. The starting point is indicated by the green square

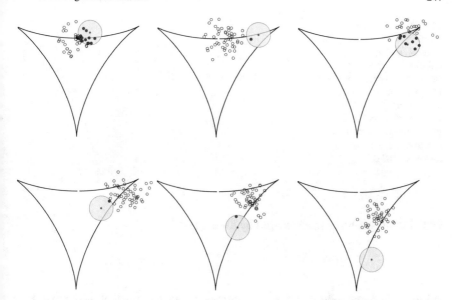

Fig. 12.3 Few snapshots of the algorithm. The red dot indicates the position of the dog, and the predictive particles are marked either by an empty dot if the likelihood weight vanishes or by a blue dot if the weight is positive. The time, in lexicographical order, is $k = 1, 2, 8, 9, 10, 11$. The data thinning, or sample impoverishment, is clearly seen here as often only one predictive particle has a positive weight. Observe that in the last step, none of the predictive particles has a positive weight, indicating that the algorithm fails

To compensate for the loss of particles, we increase the particle number to $N = 5\,000$. This time, the particle cloud survives, and we are able to estimate the position of the hunter by averaging over the particle positions,

$$\overline{x}_k = \frac{1}{N} \sum_{\ell=1}^{N} x_k^\ell.$$

Figure 12.4 shows the estimated position of the hunter. Moreover, we can use the particle sample to compute the covariance matrix of the position,

$$\mathsf{C}_k = \frac{1}{N-1} \sum_{\ell=1}^{N} (x_k^\ell - \overline{x}_k)(x_k^\ell - \overline{x}_k)^\mathsf{T}.$$

At each shown time step, around the mean position, we plot the ellipsoid corresponding to two standard deviations. This is done by first computing the symmetric singular value decomposition of the covariance matrix,

$$\mathsf{C}_k = \mathsf{U}_k \mathsf{D}_k \mathsf{U}_k^\mathsf{T},$$

Fig. 12.4 Estimated
position of the hunter (green
squares) and the uncertainty
ellipses, corresponding to
two standard deviations of
the posterior covariance
when using $N = 5000$
particles. Observe that
occasionally, the ellipses
collapse to a point due to
loss of particles

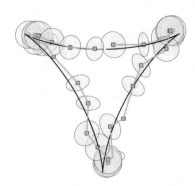

and plotting an ellipse parametrized by the angle θ,

$$s_k(\theta) = \overline{x}_k + 2\mathbf{U}_k\mathbf{D}_k^{1/2}\begin{bmatrix} \cos\theta \\ \sin\theta \end{bmatrix}, \quad 0 \le \theta \le 2\pi.$$

While most of the ellipses are of reasonable size, we observe that there are at least two time instances when the ellipses collapse to a single point, indicating that the underlying particle sample is imploded into one or few particles that are replicated.

The results highlight some of the shortcomings of the model. First, as we saw, the number of predicted particles that will be effectively discarded from the sample by being assigned zero weight can be significant, leading to an impoverishment of the sample. In fact, it may even happen that all predicted particles are discarded, causing the algorithm to stop. To avoid such catastrophic situation, we may write a less stringent likelihood model accounting for the fact that, with small probability, the dog could end up further than $d_{\max}$ from its master.[2] A likelihood conveying that sort of belief is a Gaussian model of the form,

$$\pi_{Q_1|X_1}(q_1 \mid x_1) \propto \exp\left(-\frac{1}{2(\tau d_{\max})^2}\|x_1 - q_1\|^2\right),$$

where $\tau > 0$ is a parameter expressing how strongly the belief about the distance of the dog from the master is held. Observe that, with this model, the weights w_1^ℓ defined by (12.3) never vanish, and the possibility of losing all particles is avoided.

Another assumption that we may question is whether the hunter never walks at a speed higher than $v_{\max}$, and replace the step-limited innovation with a Gaussian random walk model of the form

$$X_1 = X_0 + V_1, \quad V_1 \sim \mathcal{N}(0, (\rho\gamma_1)^2),$$

where $\rho > 0$ is a parameter controlling the expected length of the random step.

---

[2] After all, even the most trustworthy unleashed dog may occasionally wander further away than usual.

Fig. 12.5 Estimated position
of the hunter (green squares)
at discrete time instances and
the associated uncertainty
ellipses, corresponding to
two standard deviations of
the posterior covariance,
using a Gaussian likelihood
and Gaussian innovation

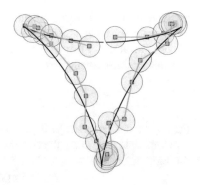

We run again a short simulation with the new model, showing the difference between the two models, using the scaling parameter $\tau = \rho = 0.4$. The number of particles is set to $N = 10\,000$. We run the algorithm with the same data as before, computing the particle mean and posterior covariances. The results are shown in Fig. 12.5. Notably, the covariance ellipses do not collapse, indicating that this model accounts better for the uncertainty in the estimate.

12.2 Sampling Importance Resampling (SIR)

Having outlined the general notion of particle filtering with the help of the preliminary example, we are ready to approach the problem in a more general setting. Before doing so, we establish the notation to be used. We begin by omitting, whenever unambiguous, the subindices specifying the random variables in probability density functions, with the understanding that the realizations will convey the information. Thus, instead of $\pi_{X|Y}(x \mid y)$, we write simply $\pi(x \mid y)$. Furthermore, we use the shorthand notation $b_{1:t}$ to indicate an accumulation of data $\{b_1, b_2, \ldots, b_t\}$ up to time t.

For the sake of definiteness, let us consider an *evolution-observation model* of random variables,

$$\begin{aligned} X_t &= F(X_{t-1}) + V_t, \\ B_t &= G(X_t) + W_t, \end{aligned} \qquad (12.4)$$

for $t = 1, 2, \ldots$. Here, the state vectors X_t describe a *discrete time stochastic process*, and the random variable V is referred to as *state noise* or *innovation process* at time t. The random variable B_t defines the *observation process*, and the variable W_t is referred to as *observation noise process* at time t. The first equation in (12.4) is the evolution model, the second one is the observation model.

In our model, we assume that F and G are known deterministic functions. Furthermore, we assume the random variables $X_j, 1 \le j \le t-1$ and V_t in the evolution model to be mutually independent, and, likewise, X_t and W_t in the observation model to be mutually independent. The evolution model, together with the independence of the innovation V_t from the past history of X, imply that the stochastic process X_t satisfies the *Markov property*,

$$\pi(x_t \mid x_{0:t-1}) = \pi(x_t \mid x_{t-1}), \tag{12.5}$$

that is, the current state X_t depends on the past $X_0, X_1, \ldots, X_{t-1}$ only through the previous state. Likewise, the current observation at time t depends on the states only through the current state,

$$\pi(b_t \mid x_{0:t}) = \pi(b_t \mid x_t). \tag{12.6}$$

The *Bayesian filtering* algorithm comprises a sequence of two-step iterations, the *prediction step* (P) and the *correction step* (C), also known as *analysis step*:

$$\pi(x_{t-1} \mid b_{1:t-1}) \xrightarrow{\text{P}} \pi(x_t \mid b_{1:t-1}) \xrightarrow{\text{C}} \pi(x_t \mid b_{1:t}),$$

where it is understood that for $t = 1$, $\pi(x_0 \mid b_{1:0}) = \pi(x_0)$, that is, the density of X_0 before any observation becomes available is simply a prior distribution. The prediction step is based on the evolution model, while the correction step follows from the observation model.

To derive the pertinent formulas, consider first the joint probability density of X_{t-1} and X_t, conditioned on the past history of observations, $B_{1:t-1}$. Using the definition of conditional densities, we write

$$\pi(x_{t-1}, x_t \mid b_{1:t-1}) = \pi(x_t \mid x_{t-1}, b_{1:t-1})\pi(x_{t-1} \mid b_{1:t-1}).$$

Next, we use the Markov properties of our model to simplify this formula. On the right-hand side, X_t depends on the past only through X_{t-1}; therefore, knowing the past observation history does not add information about X_t, implying that

$$\pi(x_t \mid x_{t-1}, b_{1:t-1}) = \pi(x_t \mid x_{t-1}).$$

In the propagation step, we are interested in how X_t depends on the past observations regardless of the value of X_{t-1}. To that end, we calculate the marginal density,

$$\pi(x_t \mid b_{1:t-1}) = \int \pi(x_{t-1}, x_t \mid b_{1:t-1})dx_{t-1}$$

$$= \int \pi(x_t \mid x_{t-1})\pi(x_{t-1} \mid b_{1:t-1})dx_{t-1}. \tag{12.7}$$

This relation, known as the *Chapman–Kolmogorov formula*, states that the predictive distribution is obtained from the previous distribution by applying an integral operator with a kernel $\pi(x_t \mid x_{t-1})$ that depends on the evolution model only.

The correction step is the result of Bayes' formula conditioned on the past observations, using the predictive distribution as the prior distribution. More precisely, we write

$$\begin{aligned}
\pi(x_t \mid b_{1:t}) &= \pi(x_t \mid b_t, b_{1:t-1}) \\
&\propto \pi(b_t \mid x_t, b_{1:t-1})\pi(x_t \mid b_{1:t-1}) \\
&= \pi(b_t \mid x_t)\pi(x_t \mid b_{1:t-1}),
\end{aligned} \tag{12.8}$$

the last equality following from the Markov property that the observation B_t depends on the past only through X_t, since the observation noise was assumed to be independent of the past. We may now combine the prediction step and the correction step into one single updating formula,

$$\pi(x_t \mid b_{1:t}) \propto \pi(b_t \mid x_t) \int \pi(x_t \mid x_{t-1})\pi(x_{t-1} \mid b_{1:t-1})dx_{t-1}. \tag{12.9}$$

This equation is the starting point for all Bayesian filtering algorithms.

We consider first the algorithm similar to what was used in the hunter-and-dog example, but with the more general evolution-observation model (12.4). Assume that at time $t - 1$, a sample drawn from the density $\pi(x_{t-1} \mid b_{1:t-1})$ with weights,

$$\mathscr{S}_{t-1} = \left\{ (x_{t-1}^1, w_{t-1}^1), (x_{t-1}^2, w_{t-1}^2), \ldots, (x_{t-1}^N, w_{t-1}^N) \right\},$$

is given. Using the idea of approximating a continuous density by a point mass density, we write

$$\pi(x_{t-1} \mid b_{1:t-1}) \approx \sum_{j=1}^{N} w_{t-1}^j \delta_{x_{t-1}^j}(x_{t-1}).$$

Substituting this approximation in (12.9) yields an approximation

$$\pi(x_t \mid b_{1:t}) \underset{\sim}{\propto} \pi(b_t \mid x_t) \sum_{j=1}^{N} w_{t-1}^j \pi(x_t \mid x_{t-1}^j), \tag{12.10}$$

where the symbol $\underset{\sim}{\propto}$ stands for approximately proportional. Next we use some of the ideas introduced in Sect. 4.7.1: For every particle x_{t-1}^j, generate a proposal particle $\widehat{x}_t^j$ drawing from $\pi(x_t \mid x_{t-1}^j)$, which is tantamount to using the propagation model,

$$\widehat{x}_t^j = F(x_{t-1}^j) + v_t^j,$$

where the innovation v_t^j is drawn independently from the probability density of V_t. In this manner, we have generated a sample of candidate particles that we then score according to how probable they make the new observation b_t. To this end, we calculate the scores and normalize them,

$$\widehat{w}_t^j = w_{t-1}^t \pi(b_t \mid \widehat{x}_t^j), \quad \widehat{w}_t^j \to \frac{\widehat{w}_t^j}{\sum_{k=1}^N \widehat{w}_t^k},$$

then resample the particles x_t^j by drawing with replacement from the set $\{\widehat{x}_t^1, \ldots, \widehat{x}_t^N\}$ using the weights $\widehat{w}_t^j$ as probabilities. In this manner, the weights of the particles in the new sample are all equal,

$$w_t^1 = \ldots w_t^N = \frac{1}{N}.$$

We are now ready to organize the different steps in the form of an algorithm.

Particle Filter Sampling Importance Resampling (PF-SIR): Assume an evolution-observation model (12.4) with a given prior density $\pi(x_0)$ of X_0.

1. **Initialize:** Draw N independent samples $x_0^1, \ldots, x_0^N$ from π_0. Set the counter $t = 1$.
2. **Repeat**: while $t \leq T$:

   (a) For each particle x_{t-1}^j, calculate $\widehat{x}_t^j = F(x_{t-1}^j) + v_t^j$, $1 \leq j \leq N$.
   (b) Compute and normalize the scores,

$$\widehat{w}_t^j = \pi(b_t \mid \widehat{x}_t^j), \quad \widehat{w}_t^j \to \frac{\widehat{w}_t^j}{\sum_{k=1}^N \widehat{w}_t^k}.$$

   (c) Generate a new sample $x_t^1, \ldots, x_t^N$ by drawing with replacement from the set $\{\widehat{x}_t^1, \ldots, \widehat{x}_t^N\}$ with probabilities $\widehat{w}_t^j$.
   (d) Advance the counter by one, $t \to t + 1$.

The PF-SIR algorithm was used in the hunter-and-dog example, with $F(x) = x$, that is, the propagation was purely a random step added to the previous position. Observe that since the weights are all equal before and after the updating in Step 2, they may be left out of the algorithm. In the algorithm, every particle x_{t-1}^j is propagated, producing one proposal particle $\widehat{x}_t^j$. This process is sometimes called *layered sampling*. It turns out that often it is better to select first the particles to be propagated, as will be shown in the next section.

12.2.1 Survival of the Fittest

The passage from the current sample at time instance $t - 1$ to the next generation sample at time instance t may be seen as an evolutionary process, with the scoring done by having the likelihood act as Darwinian selection. In the layered sampling algorithm, each parent particle generated exactly one offspring, and the offsprings were subsequently scored according to their fitness. However, mimicking the process of natural selection, one can design an algorithm in which more fit parent particles produce more offsprings than the less fit ones, and then the offsprings will be scored. The assumption is that a genetically fit parent produces offsprings that are also likely to be fit, following the popular wisdom that the apple doesn't fall far from the tree.

Our outline of the Darwinian variant of the particle filter algorithm starts with the approximate formula (12.10). Now, however, instead of layered sampling, we compute auxiliary particles without innovation,

$$\overline{x}_t^j = F(x_{t-1}^j), \quad 1 \le j \le N.$$

To check how well these particles explain the next data installment, we compute and normalize their *fitness scores*,

$$\overline{w}_t^j = w_{t-1}^j \pi(b_t \mid \overline{x}_t^j), \quad \overline{w}_t^j \to \frac{\overline{w}_t^j}{\sum_{k=1}^N \overline{w}_t^k}.$$

With the notation just introduced, we can write formula (12.10) as

$$\pi(x_t \mid b_{1:t}) \underset{\sim}{\propto} \pi(b_t \mid x_t) \sum_{j=1}^N w_{t-1}^j \pi(x_t \mid x_{t-1}^j)$$

$$\propto \sum_{j=1}^N \overline{w}_t^j \frac{\pi(b_t \mid x_t)}{\pi(b_t \mid \overline{x}_t^j)} \pi(x_t \mid x_{t-1}^j). \tag{12.11}$$

Next we select N parent particles from the auxiliary particles $\overline{x}_t^j$ for proliferation, drawing them with replacement by using the fitness scores $\overline{w}_t^j$ as probabilities. Denote the selected particles by $\overline{x}_t^{\ell_j}$, $1 \le j \le N$. Observe that it is likely to have several repetitions in this set. We then define the new particles by setting

$$x_t^j = \overline{x}_t^{\ell_j} + v_t^j,$$

where v_t^j is an independently drawn random realization of the innovation V_t. Finally, the fitness of each offspring needs to be assessed. It follows from formula (12.11) that the natural way to assign their weights is

$$w_t^j = \frac{\pi(b_t \mid x_t^j)}{\pi(b_t \mid \overline{x}_t^{\ell_j})}.$$

The sequence of operation in each survival of the fittest algorithm's iteration is illustrated schematically in Fig. 12.6.

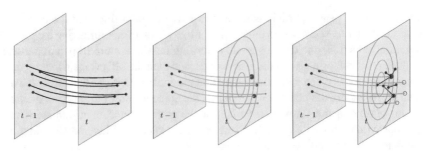

Fig. 12.6 Schematic explanation of the survival of the fittest algorithm. Left: Particles x_{t-1}^j are propagated using the deterministic forward map, $\bar{x}_t^j = F(x_{t-1}^j)$. Middle: The propagated particles are scored according to how well they explain the data in terms of the likelihood. Right: The particles generate offsprings according to their fitness score. The offsprings are then weighted according to their fitness

Before organizing the different steps in the form of an algorithm, let us examine the difference between the layered sampling algorithm and the proposed one in the light of the hunter-and-dog example. Figure 12.7 demonstrates schematically the difference. In the layered sampling algorithm (top row), each particle produces one offspring through the addition of a random step, regardless of how well, or badly, the particle is capable of explaining the next installment. As a result, several of the new offsprings may turn out to have a low weight, ending up being discarded at the following step. In the survival of the fittest algorithm, on the other hand, the particles are first scored according to their fitness, then the proliferation takes place according to the fitness score. Consequently, the new particles are typically better positioned to explain the data, leading to a higher weight and lower discarding rate. We postpone the algorithm to the next section, where a generalized version of it is derived.

12.2.2 Estimation of Static Parameters

In some cases, the deterministic forward map F depends on parameters that may be poorly known and therefore should be estimated based on the data as well. We denote by $\theta \in \mathbb{R}^k$ the vector of k real parameters, and write the evolution-observation model as

$$
\begin{aligned}
X_t &= F(X_{t-1}, \Theta) + V_t, \\
B_t &= G(X_t) + W_t,
\end{aligned}
\quad t = 1, 2, \ldots
\tag{12.12}
$$

where the parameter vector is defined as a random variable Θ with realizations θ.

To extend the PF algorithm to a parametric problem, we formally treat the random variable Θ as if it were time dependent, and use the notation Θ_t to highlight the time dependency. Observe that doing so does not mean that we believe that the parameter is changing in time, but rather that, as time marches on, our certainty about its value

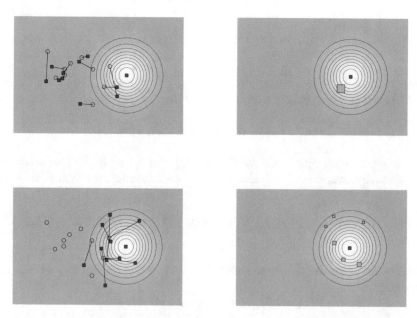

Fig. 12.7 In the left column, the old particles are denoted by empty circles, while the new particles are denoted by squares. In the top row, the layered sampling approach adds innovation to each old particle to generate new particles (blue squares). Particles are ranked according to how well they explain the data. The green squares in the right column correspond to new particles, the size of the square indicating how many times the particle is repeated in the new sample. In the bottom row, the old particles are first ranked according to their fitness to explain the data, and the more fit particles generate more offsprings

may change. As randomness in the Bayesian context represents uncertainty, each Θ_t is therefore different as a random variable. A crucial point is the choice of the innovation model for the parameter. While a random walk model is a possibility, it is not in line with the assumption that the parameter is static. A better choice is to resort to a Gaussian mixture model, introduced in Sect. 4.7.2.

We start by writing the formula (12.9) for the pair (x_t, θ_t),

$$\pi(x_t, \theta_t \mid b_{1:t}) \propto \pi(b_t \mid x_t, \theta_t)$$
$$\times \int \pi(x_t, \theta_t \mid x_{t-1}, \theta_{t-1})\pi(x_{t-1}, \theta_{t-1} \mid b_{1:t-1})dx_{t-1}d\theta_{t-1}.$$

Assuming that we have the current particle cloud,

$$\mathscr{S}_{t-1} = \left\{(x_{t-1}^1, \theta_{t-1}^1, w_{t-1}^1), \ldots, (x_{t-1}^N, \theta_{t-1}^N, w_{t-1}^N)\right\},$$

using the point mass approximation for the current density, we obtain

$$\pi(x_t, \theta_t \mid b_{1:t}) \propto \pi(b_t \mid x_t, \theta_t) \sum_{n=1}^{N} w_{t-1}^n \pi(x_t, \theta_t \mid x_{t-1}^n, \theta_{t-1}^n).$$

Next we need to write a propagation model for both x_t and θ_t. For x_t, we use the model (12.12), which defines a probability density

$$\pi(x_t \mid x_{t-1}, \theta_{t-1}) = \pi_{V_t}(x_t - F(x_{t-1}, \theta_{t-1})).$$

To define a propagation model for θ_t, recalling that x_t and θ_t are assumed to be conditionally independent, and using the Gaussian mixture approximation, we write

$$\pi(x_t, \theta_t \mid b_{1:t}) \propto \pi(b_t \mid x_t, \theta_t) \sum_{n=1}^{N} w_{t-1}^n \pi(x_t \mid x_{t-1}^n, \theta_{t-1}^n) \mathcal{N}(\theta_t \mid \overline{\theta}_{t-1}^n, h^2 C_{t-1}).$$

The auxiliary particles $\overline{\theta}_{t-1}^n$ were obtained by shifting the particles θ_{t-1}^n towards the ensemble mean,

$$\overline{\theta}_{t-1}^n = a\theta_{t-1}^n + (1-a)\overline{\theta}_{t-1}, \quad 0 < a < 1,$$

and

$$C_{t-1} = \frac{1}{N-1} \sum_{n=1}^{N} (\theta_{t-1}^n - \overline{\theta}_{t-1})(\theta_{t-1}^n - \overline{\theta}_{t-1})^{\mathsf{T}}, \quad \overline{\theta}_{t-1} = \frac{1}{N} \sum_{n=1}^{N} \theta_{t-1}^n,$$

where the parameters a and h are related by

$$a^2 + h^2 = 1.$$

Observe that in the limit as $a \to 1$, the Gaussian densities converge to point measures, i.e., there is no innovation in the parameters.

We now organize the steps of the particle filter in the form of an algorithm.

Particle Filter Survival of the Fittest (PF-SOF): Assume an evolution-observation model (12.12) with the prior densities, $\pi(x_0)$ and $\pi(\theta_0)$, of X_0 and Θ_0 given.

1. **Initialize:** Draw N independent samples $x_0^1, \ldots, x_0^N$ from π_0, and $\theta_0^1, \ldots, \theta_0^N$ from $\pi(\theta_0)$, and set the counter $t = 1$.
2. **Repeat** while $t \leq T$:

   (a) Compute the ensemble mean $\overline{\theta}_{t-1}$ and covariance C_{t-1} of the parameter vectors θ_{t-1}^n.

   (b) For each particle $(x_{t-1}^j, \theta_{t-1}^j)$, calculate $\overline{x}_t^j = F(x_{t-1}^j, \theta_{t-1}^j)$, and $\overline{\theta}_{t-1}^j = a\theta_{t-1}^j + (1-a)\overline{\theta}_{t-1}, 1 \leq j \leq N$.

(c) Compute and normalize the fitness scores,

$$g_t^j = w_{t-1}^j \pi(b_t \mid \overline{x}_t^j, \overline{\theta}_{t-1}^j), \quad g_t^j \rightarrow \frac{g_t^j}{\sum_{k=1}^{N} g_t^k}.$$

(d) Draw N indices ℓ_j with replacement from $\{1, 2, \ldots, N\}$, using the scores g_t^j as probabilities.

(e) Proliferate: Generate the new samples $(x_t^1, \theta_t^1), \ldots, (x_t^N, \theta_t^N)$,

$$x_t^j = \overline{x}_t^{\ell_j} + v_t^j, \quad \theta_t^j \sim \mathcal{N}(\theta_t \mid \overline{\theta}_t^{\ell_j}, h^2 \mathbf{C}_{t-1}),$$

where v_t^j is a realization of the innovation V_t.

(f) Calculate the weights of the particles,

$$w_t^j = \frac{\pi(b_t \mid x_t^j, \theta_t^j)}{\pi(b_t \mid \overline{x}_t^j, \overline{\theta}_t^j)}, \quad w_t^j \rightarrow \frac{w_t^j}{\sum_{k=1}^{N} w_t^k}.$$

(g) Advance the counter by one, $t \rightarrow t + 1$.

To conclude the chapter, we consider a classic tracking problem, a field in which Bayesian filtering methods have been very successful.

Example 12.2 We consider the problem of tracking an object such as a meteorite or a space module entering the atmosphere. The object's altitude is measured at given time instances by using a simple pulse radar not capable of giving an estimate of the velocity. To obtain information about the size and shape of the object, it is also of interest to assess the drag force encoded in the ballistic constant of the object.

To set up the model, we define the dynamic state variables,

$$h = h(t) = \text{altitude of the object}, \quad h > 0,$$
$$v = v(t) = \text{downwards speed}, \quad v > 0.$$

The variables are related through the basic equation of motion,

$$\frac{dh}{dt} = -v. \tag{12.13}$$

The acceleration of the object depends on two factors, the Earth's gravitational field and the drag of the atmosphere. Thus, the acceleration of the object assumes the form

$$\frac{dv}{dt} = g - g\frac{\rho(h) v^2}{2\beta}, \tag{12.14}$$

where the parameters in the equation are

$$g = \text{constant acceleration due to the gravity} = 9.81 \text{ m/s}^2,$$
$$\rho(h) = \text{altitude-dependent density of the air,}$$
$$\beta = \text{ballistic constant,} \quad \beta > 0.$$

The unknown ballistic constant depends on the shape, mass and cross-sectional area of the object, all of which are of interest but unknown. Here, we believe that the parameter β remains constant in time. Instead of including it in the filtering problem as a model parameter, we assume here a more straightforward approach, modeling it as part of the state vector, and express the belief that its value does not change in time by writing

$$\frac{d\beta}{dt} = 0. \tag{12.15}$$

For the atmospheric density, we assume that it has a known dependency on h, given by

$$\rho(h) = \gamma e^{-\eta h}, \tag{12.16}$$

where the parameters are assumed to be known, with values $\gamma = 1.754 \text{ kg/m}^3$ and $\eta = 1.39 \times 10^{-4} \text{ m}^{-1}$.

We define the state vector of unknowns as

$$x(t) = \begin{bmatrix} h(t) \\ v(t) \\ \beta \end{bmatrix}.$$

Equations (12.13)–(12.16) define a non-linear system

$$\frac{dx}{dt} = f(x) = \begin{bmatrix} -v \\ g - g\dfrac{\rho(h)\,v^2}{2\beta} \\ 0 \end{bmatrix} = \begin{bmatrix} -x_2 \\ g\left(1 - \dfrac{\rho(x_1)\,x_2^2}{2x_3}\right) \\ 0 \end{bmatrix} \tag{12.17}$$

that constitutes the basis of the propagation model.

Consider now the observation model. We assume that at times $t_j = j\Delta t$, $j = 0, 1, 2, \ldots$, the variable h is observed with some uncertainty. We write the model as

$$b_j = q^\mathsf{T} x(t_j) + \varepsilon_j, \quad q = \begin{bmatrix} 1 \\ 0 \\ 0 \end{bmatrix},$$

and we assume that the observation errors ε_j can be reasonably modeled as realizations of independent identically distributed random variables $E_j \sim \mathcal{N}(0, \sigma^2)$.

To write a discrete propagation model, marching from one observation time to the next, we use for simplicity a naïve forward Euler approximation,

$$x(t_j) \approx x(t_{j-1}) + \Delta t \, f(x(t_{j-1})).$$

In the Bayesian framework, taking into account the discretization error and possible shortcomings in the dynamic model, we define the random variables X_j that satisfy the propagation model given by

$$X_j = X_{j-1} + \Delta t f(X_{j-1}) + V_j = F(X_{j-1}) + V_j,$$

where

$$F(x) = x + \Delta t f(x),$$

and the innovations V_j are modeled for simplicity as independent identically distributed Gaussian random variables,

$$V_j \sim \mathcal{N}(0, C),$$

where the covariance matrix $C \in \mathbb{R}^{3 \times 3}$ is assumed to be diagonal. We write the observation model in terms of random variables as

$$B_j = q^{\mathsf{T}} X_j + E_j.$$

For the computed example, we start by generating the data using the system (12.17), with initial values,

$$h_0 = h^*(0) = 61\,000\,\mathrm{m}$$
$$v_0 = v^*(0) = 3\,048\,\mathrm{m/s},$$
$$\beta^* = 19\,161\,\mathrm{kg/ms}^2.$$

The data are generated by using a standard numerical ODE solver such as `ode45` of Matlab. We assume that the altitude is measured with a frequency of 10 Hz for a period of half a minute, that is, $\Delta t = 0.1\,$s and $1 \le j \le 300$. The resolution of the radar is assumed to be some hundreds of meters, which is included in the data generation by setting the standard deviation of the noise to $\sigma = 500\,$m. The time series of the data,

$$b_0, b_1, \ldots, b_T, \quad T = 300,$$

is shown in Fig. 12.8.

To estimate the state vector, we use the PF-SOF algorithm. The first step is to generate the initial particle cloud. As initial guess for the altitude, in lack of anything better, we use the first radar reading,

$$h_{\mathrm{init}} = b_0.$$

Fig. 12.8 Simulated altitude
measurements

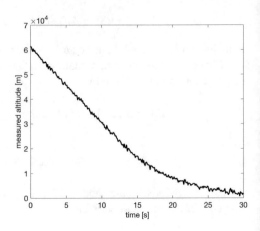

As initial guess for the velocity and ballistic coefficient, we use the values

$$v_{\text{init}} = 3\,000\,\text{m/s},$$
$$\beta_{\text{init}} = 20\,000\,\text{kg/ms}^2.$$

The number of particles is set to $N = 5\,000$. To express our uncertainty about the
initial values, the initial sample is generated by setting

$$x^\ell(t_0) = x_0^\ell = \begin{bmatrix} h_{\text{init}} \\ v_{\text{init}} \\ \beta_{\text{init}} \end{bmatrix} + \mathbf{C}^{1/2}\eta^\ell, \quad \eta^\ell \sim \mathcal{N}(0, \mathsf{I}_3),$$

where $1 \leq \ell \leq N$, and the prior covariance matrix $\mathbf{C}$ is assumed to be diagonal. The
prior standard deviations for h, v, β, or diagonal entries of $\mathbf{C}^{1/2}$, are set to

$$\sigma_h = 500\,\text{m},$$
$$\sigma_v = 200\,\text{m/s},$$
$$\sigma_\beta = 1\,500\,\text{kg/ms}^2.$$

Furthermore, we assign equal weights $w_0^\ell = 1/N$ to all particles.

Next we outline the details of the algorithm. At each time instance t_j, $j =
1, \ldots, T$,

1. Propagate each particle x_{j-1}^ℓ using the propagation model F,

$$\widehat{x}_j^\ell = F(x_{j-1}^\ell), \quad 1 \leq \ell \leq N.$$

2. For every candidate particle, compute the fitness weight,

$$\widehat{w}_j^\ell = w_{j-1}^\ell \exp\left(-\frac{1}{2\sigma^2}|q^\mathsf{T}\widehat{x}_j^\ell - b_j|^2\right),$$

and normalize them so as to have the sum equal to one.

3. Using the fitness weights as probabilities, draw with replacement N indices $k_\ell \in \{1, 2, \ldots, n\}$, and generate the new particles by adding innovation,

$$x_j^\ell = \widehat{x}^{k_\ell} + D^{1/2}\xi^\ell, \quad \xi^\ell \sim \mathcal{N}(0, I_3),$$

where $D \in \mathbb{R}^{3\times3}$ is the covariance matrix of the innovation. We assume that the innovation covariance matrix is diagonal, and the standard deviations of the innovation, or the diagonal entries of $D^{1/2}$, are

$$\sigma_h = 100\,\text{m},$$
$$\sigma_v = 100\,\text{m/s},$$
$$\sigma_\beta = 5\,\text{kg/ms}^2.$$

4. Update the weights, by computing

$$w_j^\ell = \exp\left(-\frac{1}{2\sigma^2}\left(|(q^\mathsf{T}x_j^\ell - b_j|^2 - |q^\mathsf{T}\widehat{x}_j^{k_\ell} - b_j|^2)\right)\right),$$

followed by a normalization step.

We run the described algorithm, and for each time step, estimate the particle mean and standard deviation. Figure 12.9 shows the results, plotted against the generative target values. We see that the algorithm is capable of identifying the velocity profile rather well. The ballistic constant estimate shows some drift towards the end of the estimation period, possibly due to the diminishing sensitivity of the data to the parameter towards the end of the period, when the velocity decreases.

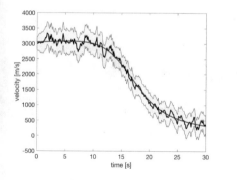 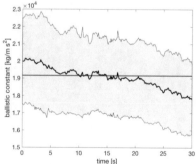

Fig. 12.9 The average estimated velocity (left panel) and ballistic constant (right panel) are the black curve, while the red curves are the values in the generative model. The envelopes correspond to one standard deviation uncertainty

Notes and Comments

As in the case of Markov chain Monte Carlo, the literature on particle filtering methods is rather extensive. We refer to [28] for a nice overview, and to the article [53] in the collection [29] which contains a number of further relevant contributions.

The tracking problem at the end of the chapter originates from the book [63].

Chapter 13
Bayesian Filtering for Gaussian Densities

In the previous chapter, we considered dynamic inverse problems where the posterior density is updated sequentially as new observations become available. The particle filter approach is fully general and does not assume anything particular about the probability densities, as they were approximated by particle-based point mass distributions. However, if parametric forms of the distributions are known, or if the distributions can be approximated by parametric ones, computationally efficient methods for Bayesian filtering are available. In this section, we review methods that make use of Gaussian densities which, being completely characterized by the mean and the covariance, allow a natural parametric representation. In particular, we derive the classical Kalman filtering algorithm, and discuss some generalizations of it.

13.1 Kalman Filtering

Consider an evolution-observation model where we have a stochastic process $X_0, X_1, \ldots$ taking on values in $\mathbb{R}^n$, and an observation process $B_1, B_2, \ldots$, taking on values in $\mathbb{R}^m$, such that

$$X_t = \mathsf{F} X_{t-1} + V_t, \tag{13.1}$$

$$B_t = \mathsf{A} X_t + W_t, \tag{13.2}$$

where $t = 1, 2, \ldots$, and $\mathsf{F} \in \mathbb{R}^{n \times n}$ and $\mathsf{A} \in \mathbb{R}^{m \times n}$. Moreover, we assume that

1. $X_0 \sim \mathcal{N}(\overline{x}_0, \mathsf{D}_0)$;
2. $V_t \sim \mathcal{N}(0, \Gamma_t)$, $W_t \sim \mathcal{N}(0, \Sigma_t)$;
3. The random variables X_0, V_t and W_t are mutually independent.

As in the case of particle filtering, our goal is to compute the conditional probability densities $\pi_{X_t | B_{1:t}}(x_t \mid b_{1:t})$, where we use the shorthand notation

© The Author(s), under exclusive license to Springer Nature Switzerland AG 2023
D. Calvetti and E. Somersalo, *Bayesian Scientific Computing*, Applied Mathematical Sciences 215, https://doi.org/10.1007/978-3-031-23824-6_13

$$B_{1:t} = \{B_1, B_2, \ldots, B_t\},$$

to indicate the accumulation of the observations up to the current time t.

A well-known result concerning Gaussian random variables states that *any linear combination of Gaussian random variables is also a Gaussian random variable.* This implies that if a matrix is applied to a Gaussian random vector, the outcome is also Gaussian, and the sum of two Gaussian random vectors is a Gaussian random vector. The proof of these statements is not difficult, and is left as an exercise. In particular, it follows from the form of the system (13.1)–(13.2) and the assumptions about the stochastic processes that the vectors X_t and B_t are normally distributed for all t.

In the derivation of the updating formulas for the conditional densities, we observe that since a Gaussian density is completely specified by its mean and its covariance matrix, it suffices to find an updating formula for the mean and the covariance matrices. As in the case of the particle filter, we examine the prediction step and the correction step separately.

We start with the prediction step and the evolution Eq. (13.1). Assume that the Gaussian densities,

$$X_{t-1} \mid B_{t-1} \sim \mathcal{N}(\overline{x}_{t-1}, \mathsf{D}_{t-1}), \quad V_t \sim \mathcal{N}(0, \Gamma_t),$$

are given, and that V_t is independent of the past history. Based on this information, we can calculate the mean X_t, conditioned on $B_{1:t-1}$,

$$\mathrm{E}(X_t \mid B_{1:t-1}) = \mathsf{F}\mathrm{E}(X_{t-1} \mid B_{1:t-1}) + \mathrm{E}(V_t \mid B_{1:t-1})$$
$$= \mathsf{F}\overline{x}_{t-1}.$$

To calculate the covariance, observe that from the evolution equation,

$$X_t - \mathrm{E}(X_t \mid B_{1:t-1}) = \mathsf{F}(X_{t-1} - \overline{x}_{t-1}) + V_t,$$

hence, by the independency of X_{t-1} and V_t,

$$\mathrm{E}\big((X_t - \mathrm{E}(X_t))(X_t - \mathrm{E}(X_t))^{\mathsf{T}} \mid B_{1:t-1}\big)$$
$$= \mathsf{F}\mathrm{E}\big((X_{t-1} - \overline{x}_{t-1})(X_{t-1} - \overline{x}_{t-1})^{\mathsf{T}} \mid B_{1:t-1}\big)\mathsf{F}^{\mathsf{T}} + \mathrm{E}(V_t V_t^{\mathsf{T}})$$
$$= \mathsf{F}\mathsf{D}_{t-1}\mathsf{F}^{\mathsf{T}} + \Gamma_t.$$

We have therefore established that, based on the evolution model alone,

$$\pi_{X_t \mid B_{1:t-1}}(x_t \mid b_{1:t-1}) = \mathcal{N}(x_t \mid \widehat{x}_t, \widehat{D}_t), \tag{13.3}$$

where

$$\widehat{x}_t = F\overline{x}_{t-1}, \tag{13.4}$$
$$\widehat{D}_t = FD_{t-1}F^T + \Gamma_t. \tag{13.5}$$

Observe that the above formulas do not give any indication how the Gaussian density depends on the earlier observations, because this information is implicitly contained in the mean and covariance at the previous time step.

We now consider the correction step, as the new observation $B_t = b_t$ arrives. According to Bayes' formula,

$$\pi_{X_t|B_t}(x_t \mid b_t) \propto \pi_{X_t}(x_t)\pi_{B_t|X_t}(b_t \mid x_t).$$

To condition the density of X_t also on all previous information $B_{1:t-1}$, observe that the likelihood of B_t depends only on X_t. Therefore, as in the particle filtering problem, we can write

$$\pi_{X_t|B_{1:t}}(x_t \mid b_{1:t}) = \pi_{X_t|B_t,B_{1:t-1}}(x_t \mid b_t, b_{1:t-1})$$
$$\propto \pi_{X_t|B_{1:t-1}}(x_t \mid b_{1:t-1})\pi_{B_t|X_t,B_{1:t-1}}(b_t \mid x_t, b_{1:t-1})$$
$$= \pi_{X_t|B_{1:t-1}}(x_t \mid b_{1:t-1})\pi_{B_t|X_t}(b_t \mid x_t).$$

In other words, we consider the conditional density (13.3) as a prior, and derive the posterior density given the new observation. Since everything is Gaussian, we can use the explicit expressions for the posterior mean and covariance derived in Chap. 8. In particular, from (8.11) and (8.10), we have

$$\pi_{X_t|B_{1:t}}(x_t \mid b_{1:t}) = \mathcal{N}(x_t \mid \overline{x}_t, D_t),$$

where

$$\overline{x}_t = \widehat{x}_t + \widehat{D}_t A^T (A\widehat{D}_t A^T + \Sigma_t)^{-1}(b_t - A\widehat{x}_t),$$
$$D_t = \widehat{D}_t - \widehat{D}_t A^T (A\widehat{D}_t A^T + \Sigma_t)^{-1}A\widehat{D}_t.$$

This completes the correction step.

In the classical literature, the vector

$$\Delta_t = b_t - A\widehat{x}_t,$$

expressing how much $\widehat{x}_t$ fails to predict the next observation is referred to as *measurement residual*, and the matrix

$$K_t = \widehat{D}_t A^T (A\widehat{D}_t A^T + \Sigma_t)^{-1}$$

is referred to as the *Kalman gain* matrix.

We are ready now to present the sequential update process in the form of an algorithm.

Kalman Filtering algorithm: Given $\bar{x}_0$, the covariance matrices D_0, Γ_t, Σ_t; the matrices F and A; and the data vectors $b_1, \ldots, b_T$.

1. **Initialize:** Set the counter $t = 1$.
2. **Iterate:** While $t < T$,

   (a) **Prediction:** Update the predicted mean and covariance,

   $$\widehat{x}_t = F\bar{x}_{t-1}, \tag{13.6}$$
   $$\widehat{D}_t = FD_{t-1}F^T + \Gamma_t. \tag{13.7}$$

   (b) **Correction:** On arrival of new data b_t, update the posterior mean and covariance,

   $$\bar{x}_t = \widehat{x}_t + \widehat{D}_t A^T \left(A\widehat{D}_t A^T + \Sigma_t\right)^{-1} (b_t - A\widehat{x}_t),$$
   $$D_t = \widehat{D}_t - \widehat{D}_t A^T \left(A\widehat{D}_t A^T + \Sigma_t\right)^{-1} A\widehat{D}_t.$$

   (c) Advance the counter $t \to t + 1$.

13.2 The Best of Two Worlds: Ensemble Kalman Filtering

The appeal of the Kalman filtering algorithm is its straightforward linear algebraic nature, and its drawback is that the assumptions about the model are quite restrictive and not justified in many real-world problems. Particle methods, on the other hand, are very intuitive as far as the particle propagation is concerned, but the correction or analysis step can be quite challenging, and making a particle filtering algorithms work can require a lot of tuning and testing.

Another issue with particle filtering algorithms is the sample size. In complex high-dimensional problems, the number of particles needed for a reliable uncertainty analysis tends to be large, which may be a real bottleneck in applications such as weather prediction, where the real-time nature of the process poses strict constraints on computing times.

The idea behind the algorithm that we present in this section is intuitive: Because of the conceptual clarity of the particle propagation, the prediction step is done as in particle filtering. The key component of the correction step, on the other hand, is Bayes' formula, stating that the posterior is the product of the prior and the likelihood, hence there are two sources for the uncertainty in the posterior density, namely, the uncertainty in the prior and the uncertainty in the likelihood. In the case of Gaussian densities, we can analyze the posterior uncertainty by perturbing the mean of the prior and the data separately, and analyzing the joint effect of these perturbations on the posterior. The technique can be extended approximately to densities not too far from Gaussians. In this manner, we arrive at an algorithm combining the best of the particle and the parametric worlds.

In the following, we formalize the ideas outlined above, starting with the prediction step. Consider the evolution-observation model

$$X_t = F(X_{t-1}) + V_t, \atop B_t = G(X_t) + W_t, \quad t = 1, 2, \ldots \tag{13.8}$$

where

$$V_t \sim \mathcal{N}(0, \Gamma_t), \quad W_t \sim \mathcal{N}(0, \Sigma_t). \tag{13.9}$$

Consider a particle sample,

$$\mathcal{S}_{t-1} = \left\{ x_{t-1}^1, x_{t-1}^2, \ldots, x_{t-1}^N \right\}$$

drawn from the conditional density $\pi_{X_{t-1}|B_{1:t-1}}$, where the particles have all equal weight $1/N$. To update the sample according to the evolution model, we perform the prediction step as in the particle filtering algorithm. For simplicity, we follow the layered sampling algorithm, and generate the prediction sample

$$\widehat{\mathcal{S}}_t = \left\{ \widehat{x}_t^1, \widehat{x}_t^2, \ldots, \widehat{x}_t^N \right\},$$

where

$$\widehat{x}_t^j = F(x_{t-1}^j) + v_t^j, \quad v_t^j \sim \mathcal{N}(0, \Gamma_t), \quad 1 \le j \le N.$$

Next we calculate the predictive mean and covariance,

$$\widehat{x}_t = \frac{1}{N} \sum_{j=1}^{N} \widehat{x}_t^j, \quad \widehat{D}_t = \frac{1}{N-1} \sum_{j=1}^{N} (\widehat{x}_t^j - \widehat{x}_t)(\widehat{x}_t^j - \widehat{x}_t)^{\mathsf{T}}.$$

At this point, if the predictive sample represents a Gaussian distribution, and if the observation model is linear, we can continue as in Kalman filtering, using the updating formulas to get a posterior distribution. In the case where at least one of these conditions is not satisfied, we introduce the following modification, inspired by the solution of linear inverse problems.

Consider a linear inverse problem with a Gaussian prior and Gaussian additive noise,

$$B = AX + W, \quad W \sim \mathcal{N}(0, \Sigma), \quad x \sim \mathcal{N}(\widehat{x}, \widehat{D}). \tag{13.10}$$

The posterior density, which is proportional to the product of the prior and the likelihood, is of the form

$$\pi_{X|B}(x \mid b) \propto \exp\left(-\frac{1}{2}\mathscr{E}(x, b)\right),$$

where the *Gibbs energy* is given by

$$\mathscr{E}(x, b) = \|Ax - b\|_{\Sigma}^2 + \|x - \widehat{x}\|_{\widehat{D}}^2.$$

Consider now the following sampling scheme, called *Randomize, then optimize (RTO) algorithm*, to explore the posterior density.

1. For $1 \le j \le N$, draw two mutually independent realizations of Gaussian random variables,

$$\eta^j \sim \mathcal{N}(0, \Sigma), \quad \nu^j \sim \mathcal{N}(0, \widehat{D}), \tag{13.11}$$

   and perturb the data and prior mean by defining

$$b^j = b + \eta^j, \quad \widehat{x}^j = \widehat{x} + \nu^j. \tag{13.12}$$

2. Define
$$x^j = \operatorname{argmin}\{\|Ax - b^j\|_{\Sigma}^2 + \|x - \widehat{x}^j\|_{\widehat{D}}^2\}. \tag{13.13}$$

It turns out that the algorithm above produces a sample

$$\mathscr{S} = \{x^1, \ldots, x^N\},$$

which is distributed according to the posterior density. To prove it, consider the symmetric decompositions of the precision matrices,

$$\Sigma^{-1} = S^T S, \quad \widehat{D}^{-1} = L^T L,$$

and write the Gibbs energy as

$$\mathscr{E}(x, b) = \|SAx - Sb\|^2 + \|Lx - L\widehat{x}\|^2 = \left\|\begin{bmatrix} SA \\ L \end{bmatrix} x - \begin{bmatrix} Sb \\ L\widehat{x} \end{bmatrix}\right\|^2.$$

The minimizer of the Gibbs energy is the least squares solution of the linear system

$$\begin{bmatrix} SA \\ L \end{bmatrix} x = \begin{bmatrix} Sb \\ L\widehat{x} \end{bmatrix},$$

and also the solution of the normal equations,

$$((SA)^T SA + L^T L)x = (SA)^T Sb + L^T L\widehat{x},$$

which can be expressed in terms of the covariance matrices as

$$(A^\mathsf{T}\Sigma^{-1}A + \widehat{D}^{-1})x = A^\mathsf{T}\Sigma^{-1}b + \widehat{D}^{-1}\widehat{x}.$$

Not surprisingly, the solution is the posterior mean estimate,

$$\bar{x} = (A^\mathsf{T}\Sigma^{-1}A + \widehat{D}^{-1})^{-1}(A^\mathsf{T}\Sigma^{-1}b + \widehat{D}^{-1}\widehat{x}). \tag{13.14}$$

Now, instead of solving for the posterior mean, we first perturb b and $\widehat{x}$ according to (13.12) and (13.11): Let ν and η be independent Gaussian random variables,

$$\eta \sim \mathcal{N}(0, \Sigma), \quad \nu \sim \mathcal{N}(0, \widehat{D}),$$

and consider the normal equations,

$$(A^\mathsf{T}\Sigma^{-1}A + \widehat{D}^{-1})X = A^\mathsf{T}\Sigma^{-1}(b + \eta) + \widehat{D}^{-1}(\widehat{x} + \nu). \tag{13.15}$$

The solution X to (13.15) is a random variable, with mean

$$E(X) = \bar{x},$$

as can be readily verified by computing the expectation on both sides of Eq. (13.15), yielding (13.14). Furthermore, by subtracting Eq. (13.14) from (13.15) side by side, we find that

$$(A^\mathsf{T}\Sigma^{-1}A + \widehat{D}^{-1})(X - \bar{x}) = A^\mathsf{T}\Sigma^{-1}\eta + \widehat{D}^{-1}\nu.$$

To find the covariance of X, we first solve the above equation for $X - \bar{x}$, obtaining

$$\begin{aligned} X - \bar{x} &= (A^\mathsf{T}\Sigma^{-1}A + \widehat{D}^{-1})^{-1}(A^\mathsf{T}\Sigma^{-1}\eta + \widehat{D}^{-1}\nu) \\ &= D(A^\mathsf{T}\Sigma^{-1}\eta + \widehat{D}^{-1}\nu), \end{aligned}$$

where

$$D = (A^\mathsf{T}\Sigma^{-1}A + \widehat{D}^{-1})^{-1}.$$

Then, because of the mutual independency of η and ν, the covariance matrix of X turns out to be

$$\begin{aligned} E((X - \bar{x})(X - \bar{x})^\mathsf{T}) &= DE\left\{(A^\mathsf{T}\Sigma^{-1}\eta + \widehat{D}^{-1}\nu)(A^\mathsf{T}\Sigma^{-1}\eta + \widehat{D}^{-1}\nu)^\mathsf{T}\right\}D^\mathsf{T} \\ &= D\left\{A^\mathsf{T}\Sigma^{-1}E(\eta\eta^\mathsf{T})\Sigma^{-1}A + \widehat{D}^{-1}E(\nu\nu^\mathsf{T})\widehat{D}^{-1}\right\}D \\ &= D(A^\mathsf{T}\Sigma^{-1}\Sigma\Sigma^{-1}A + \widehat{D}^{-1}\widehat{D}\widehat{D}^{-1})D \\ &= D(A^\mathsf{T}\Sigma^{-1}A + \widehat{D}^{-1})D \\ &= D. \end{aligned}$$

In conclusion,

$$X \sim \mathcal{N}(\overline{x}, \mathsf{D}),$$

which is exactly the posterior distribution corresponding to the problem (13.10), as claimed.

This result motivates an algorithm where the correction step of the particle filtering algorithm is replaced by an RTO step, which, in the non-linear and non-Gaussian case, gives a sample approximation of the posterior, while in the linear Gaussian case produces an exact sample from the posterior.

We summarize the procedure in the following *Ensemble Kalman Filtering (EnKF)* algorithm.

Ensemble Kalman Filtering algorithm: Given the prior density π_{X_0}, the evolution-observation model (13.8) with covariance matrices (13.9), and the data vectors $b_1, \ldots, b_T$.

1. **Initialize:** Draw a sample $\{x_0^1, \ldots, x_0^N\}$ from the prior density π_{X_0} by independent sampling. Set the counter $t = 1$.
2. **Iterate:** While $t < T$,

   (a) **Prediction:** Propagate the sample to get the predictive sample

   $$\widehat{x}_t^j = F(x_{t-1}^j) + v_t^j, \quad v_t^j \sim \mathcal{N}(0, \Gamma_t).$$

   (b) Compute an estimate of the predictive covariance matrix,

   $$\widehat{\mathsf{D}}_t = \frac{1}{N-1} \sum_{j=1}^N (\widehat{x}_t^j - \widehat{x}_t)(\widehat{x}_t^j - \widehat{x}_t)^\mathsf{T} + \alpha \mathsf{I}_n,$$

   where

   $$\widehat{x}_t = \frac{1}{N} \sum_{j=1}^N \widehat{x}_t^j,$$

   and $\alpha > 0$ is a small variance inflation parameter.

   (c) **Correction:** On arrival of new data, perturb the data, generating an artificial sample of data

   $$b_t^j = b_t + \eta_t^j, \quad \eta_t^j \sim \mathcal{N}(0, \Sigma_t), \quad 1 \le j \le N.$$

   (d) For every j, solve the minimization problem

   $$x_t^j = \mathrm{argmin}\big\{ \|b_t^j - G(x_t)\|_{\Sigma_t}^2 + \|x_t - \widehat{x}_t^j\|_{\widehat{\mathsf{D}}_t}^2 \big\}.$$

   (e) Advance the counter $t \rightarrow t + 1$.

The algorithm deserves a few comments. First, observe that in step (b), a small multiple of the identity matrix is added to the sample-based estimate of the covariance matrix to ensure that the resulting covariance is positive definite, which is not guaranteed when the sample is used. This is often the case for high-dimensional problems, where modest sample sizes yield only a positive semi-definite matrix.

A second comment is that we add no perturbation to the propagated particles, since we tacitly assume that they are already distributed around the ensemble mean according to the empirical covariance.

The final comment concerns step (d). If the observation function G is non-linear, a non-linear optimizer must be used to compute the minimizer, while if the mapping G is linear, $G(x) = Ax$ for some matrix A, the problem reduces to a least squares problem, and the formula for the minimizer is essentially that of the posterior mean. In fact, in this special case,

$$b_t = Ax_t + w_t,$$

and the analysis step reduces to a least squares problem with the explicit solution,

$$x_t^j = \widehat{x}_t^j + K_t \left(b_t^j - A\widehat{x}_t^j \right), \qquad j = 1, 2, \ldots, N, \qquad (13.16)$$

where K_t is the Kalman gain matrix

$$K_t = \widehat{D}_t A^\mathsf{T} \left(A\widehat{D}_t A^\mathsf{T} + \Sigma_t \right)^{-1}.$$

13.2.1 Adding Unknown Parameters

As in the particle filtering algorithm, the forward propagation model F may depend on unknown parameters that need to be estimated simultaneously with the state. Typically, we may have a propagation model

$$x_t = F(x_{t-1}, \theta) + V_t,$$

where $\theta \in \mathbb{R}^k$ is a poorly known parameter vector. Even if we assume that the parameter θ is static, we model it as a the time-dependent random variable Θ_t, the time dependency reflecting only the change in our information about its value, and introduce a propagation model for an extended variable Z_t,

$$Z_t = \begin{bmatrix} X_t \\ \Theta_t \end{bmatrix} = \begin{bmatrix} F(X_{t-1}, \Theta_{t-1}) \\ \Theta_{t-1} \end{bmatrix} = \mathscr{F}(X_{t-1}).$$

To understand how the EnKF algorithm learns about the parameter values, consider the special case in which the data consist of direct observation of some components of the state vector X_t, as is sometimes the case in chemical kinetics, where the concentrations of one or two substances that participate in the reaction are measured. Thus, the observation model is given by

$$B_t = PX_t + W_t = \begin{bmatrix} P & O_{m \times k} \end{bmatrix} Z_t + W_t,$$

where P is a projection matrix onto the observed components, obtained by deleting all rows that correspond to non-observed components from the identity matrix I_n. By partitioning the predictive covariance matrix $\widehat{D}_t \in \mathbb{R}^{(n+k) \times (n+k)}$ of the extended variable Z_t as

$$\widehat{D}_t = \begin{bmatrix} \widehat{D}_{xx} & \widehat{D}_{x\theta} \\ \widehat{D}_{\theta x} & \widehat{D}_{\theta\theta} \end{bmatrix} \in \mathbb{R}^{(n+k) \times (n+k)},$$

the Kalman gain matrix becomes

$$K_t = \begin{bmatrix} \widehat{D}_{xx} \\ \widehat{D}_{\theta x} \end{bmatrix} P^T (P\widehat{D}_{xx}P^T + \Sigma_t)^{-1}.$$

The block partitioning of the Kalman gain reveals that the algorithm learns about the parameter vector θ through the correlation between the variables X_t and Θ_t, since the updating of the components corresponding to Θ_t is nonzero only if the parameter is correlated with the state vector.

We close this chapter with a computed example elucidating the ideas of the EnKF algorithm.

Example 13.1 Consider a simple example of a reversible chemical reaction,

$$A \underset{k_2}{\overset{k_1}{\rightleftharpoons}} B,$$

where k_1 and k_2 are reaction rates. Assuming that the reactions satisfy the mass action model, the concentrations c_1 and c_2 of the substances A and B, respectively, satisfy the system

$$\frac{dc_1}{dt} = -k_1 c_1 + k_2 c_2,$$

$$\frac{dc_2}{dt} = -k_1 c_1 - k_2 c_2.$$

Given the initial values

$$c_1(0) = c_A, \quad c_2(0) = c_B,$$

we may write the solution to the initial value problem explicitly as

$$c_1(t; k_1, k_2) = \alpha + \beta e^{-t/\tau},$$
$$c_2(t; k_1, k_2) = \alpha\delta - \beta e^{-t/\tau},$$

where the time constant τ is given by

$$\tau = \frac{1}{k_1 + k_2},$$

and the coefficients α and β are given by

$$\alpha = \frac{c_A + c_B}{1 + \delta}, \quad \beta = \frac{\delta c_A - c_B}{1 + \delta}, \quad \delta = \frac{k_1}{k_2}.$$

We consider the following problem: Assuming that the initial values c_A and c_B are measured with a given accuracy, we want to estimate the parameters k_1 and k_2 from noisy observations of the concentration of the substance A,

$$b_j = c_1(t_j; k_1, k_2) + e_j, \quad 1 \le j \le m,$$

where $0 < t_1 < \ldots < t_m$. For simplicity, we will assume that e_j is a realization of a random variable $E_j \sim \mathcal{N}(0, \sigma^2)$, and that the variables E_j are mutually independent.

At first, we assume that both concentrations c_A and c_B are measured with the same precision as the data. Furthermore, we assume that the parameters satisfy the a priori bound constraints,

$$k_{j,\min} \le k_j \le k_{j,\max}, \quad j = 1, 2.$$

We introduce the combined state and parameters vector

$$x = \begin{bmatrix} x_1 \\ x_2 \\ x_3 \\ x_4 \end{bmatrix} = \begin{bmatrix} c_1 \\ c_2 \\ k_1 \\ k_2 \end{bmatrix},$$

and define the propagation model

$$\frac{dx}{dt} = F(x) = \begin{bmatrix} -x_3 x_1 + x_4 x_2 \\ x_3 x_1 - x_4 x_2 \\ 0 \\ 0 \end{bmatrix}. \tag{13.17}$$

The observation model in terms of x is

$$b_j = Ax(t_j) + e_j, \quad A = \begin{bmatrix} 1 & 0 & 0 & 0 \end{bmatrix}.$$

We generate the data using the generative values

$$c_A^* = 2, \quad c_B^* = 1, \quad k_1^* = 2, \quad k_2^* = 0.5$$

Fig. 13.1 The concentration time courses of the two substances corresponding to the generative model. The simulated data are indicated in the figure by the dots. The noise level, given in terms of the standard deviation of the normal distribution, is $\sigma = 0.05$

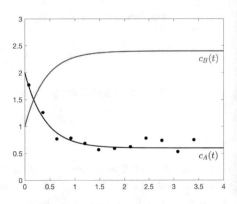

at $m = 12$ time instances, and corrupted by additive Gaussian scaled white noise with $\sigma = 0.05$. The noiseless curves $c_1(t)$ and $c_2(t)$ with the noisy data are shown in Fig. 13.1.

To start the EnKF algorithm, we choose the ensemble size to be $N = 5\,000$, and generate the initial particle ensemble as follows: Assume that the measured initial values are

$$c_A = 2.1, \quad c_B = 0.9.$$

We define the initial particles x_0^ℓ component-wise,

$$x_{0,1}^\ell = c_A + \sigma w_1^\ell, \quad x_{0,2}^\ell = c_B + \sigma w_2^\ell,$$

where

$$w_1^\ell, w_2^\ell \sim \mathcal{N}(0, 1),$$

reflecting the fact that the measurement precision is the same as in the process of collecting data. Further, we set

$$x_{0,3}^\ell = k_{1,\min} + \left(k_{1,\max} - k_{1,\min}\right)w_3^\ell, \quad x_{0,4}^\ell = k_{2,\min} + \left(k_{2,\max} - k_{2,\min}\right)w_4^\ell,$$

where

$$w_3^\ell, w_4^\ell \sim \text{Uniform}\big([0, 1]\big),$$

and assume that

$$k_{1,\min} = 0.5, \quad k_{1,\max} = 3, \quad k_{2,\min} = 0.1, \quad k_{2,\max} = 1.5.$$

Consider now the EnKF iteration. In order to keep the notations as light as possible, we adopt the following convention: In the propagation step, x^ℓ refers to the ℓth particle at time t_{j-1}, and $\widehat{x}^\ell$ to the prediction at time t_j. Likewise, in the correction step, we write for brevity $\widehat{x}^{\ell}$ for $\widehat{x}_{j}^{\ell}$, and the new particles after the correction step overwrite the old x^{ℓ}.

With this notation, we denote the current particle ensemble at time t_{j-1} by

$$\{x^1, x^2, \ldots, x^N\}, \quad x^{\ell} \in \mathbb{R}^4.$$

For the prediction step, we propagate the ensemble by solving the system (13.17) with initial values $x(t_{i-1}) = x^{\ell}$, either by using a numerical integrator or the analytic expressions,

$$\begin{aligned}
\widehat{x}_1^{\ell} &= \alpha^{\ell} + \beta^{\ell} e^{-(t_i - t_{i-1})/\tau^{\ell}}, \\
\widehat{x}_2^{\ell} &= \alpha^j \delta^{\ell} - \beta^{\ell} e^{-(t_i - t_{i-1})/\tau^{\ell}}, \\
\widehat{x}_3^{\ell} &= x_3^{\ell}, \\
\widehat{x}_4^{\ell} &= x_4^{\ell},
\end{aligned}$$

where

$$\tau^{\ell} = \frac{1}{x_3^{\ell} + x_4^{\ell}}, \quad \delta^{\ell} = \frac{x_3^{\ell}}{x_4^{\ell}},$$

and

$$\alpha^{\ell} = \frac{x_1^{\ell} + x_2^{\ell}}{1 + \delta^{\ell}}, \quad \beta^{\ell} = \frac{\delta^{\ell} x_1^{\ell} - x_2^{\ell}}{1 + \delta^{\ell}}.$$

This way, we obtain the predictive ensemble

$$\{\widehat{x}^1, \widehat{x}^2, \ldots, \widehat{x}^N\}, \quad \widehat{x}^{\ell} \in \mathbb{R}^4. \tag{13.18}$$

Observe that here we trust the forward model, and therefore do not add any random innovation.

At $t = t_j$, the new data $b = b_j$ arrives, and the predictions need to be updated. To this end, we compute the ensemble mean $\widehat{x}$ and covariance $\widehat{D} \in \mathbb{R}^{4 \times 4}$ of the predictive ensemble (13.18). We generate the artificial data cloud,

$$b^{\ell} = b + \sigma v^{\ell}, \quad v^{\ell} \sim \mathcal{N}(0, 1),$$

and update the predictive cloud. Since our observation model is linear, the updating can be made by using the formula (13.16). In the present case, the observation is a scalar, and therefore

$$(\mathsf{A}^{\mathsf{T}} \widehat{\mathsf{D}} \mathsf{A} + \Sigma)^{-1} = \frac{1}{\widehat{d}_{11} + \sigma^2}, \quad \widehat{\mathsf{D}} \mathsf{A}^{\mathsf{T}} = \widehat{d}_1,$$

where $\widehat{d}_1 \in \mathbb{R}^4$ is the first column of the matrix $\widehat{D}$ and $\widehat{d}_{11}$ is its first component. Therefore, the updating simplifies to

$$x^\ell = \widehat{x}^\ell + \frac{b^\ell - \widehat{x}_1^\ell}{\widehat{d}_{11} + \sigma^2}\widehat{d}_1.$$

In Fig. 13.2, the results have been summarized by plotting the ensemble predictive envelopes: At each observation time t_j, we compute the ensemble mean, and for a given percentage p, we find an interval containing $p\%$ of the particle values, simply by discarding $(100 - p)/2$ largest and smallest particle values. For visualization purposes, the upper and lower bounds of the intervals are connected to form an envelope. The figure shows the envelopes of $p = 90$ and $p = 75$.

The results show that the EnKF algorithm manages to identify the concentration of B with an accuracy in line with the observation error. Furthermore, we notice that after the system reaches near equilibrium, the parameter estimates do not improve, and, in fact, their estimation error seems to increase. One explanation for this may be that after the system reaches equilibrium, the solution $c_1(t)$ loses exponentially fast its dependency on the time constant τ, depending only on the ratio δ of the reaction rates. A disturbing feature in the results is the negativity of some of the particle components corresponding to the parameter k_2. The negative values are the result of applying the Kalman filtering formula in the correction step. One way to avoid negative values for a one that is known to be positive is to define the state vector using a logarithmic transformation, and define

$$x = \begin{bmatrix} x_1 \\ x_2 \\ x_3 \\ x_4 \end{bmatrix} = \begin{bmatrix} c_1 \\ c_2 \\ \log k_1 \\ \log k_2 \end{bmatrix}.$$

Figure 13.3 shows the results with this modification. The negative particle values are eliminated, and the estimated values are closer to the generative values.

Finally, recall that in the correction step of the EnKF algorithm, the unobserved components are updated from the data on the basis of the correlation between the observed and unobserved quantities. Therefore, it is instructive to look at the correlation between these variables by using the particle sample. Figure 13.4 displays scatter plots of the three unobserved quantities against the observed one. These plots give a good idea how uncertainties in the observations propagate to uncertainties in the unobserved quantities.

Notes and Comments

The idea of Kalman filtering is usually attributed to Rudolf Kálmán [50] and Ruslan Stratonovich [73], although the full history is much richer. The word "filtering" refers to the original use of the methodology for signal denoising. The engineering literature on Kalman filtering is vast due to its extensive use in signal processing.

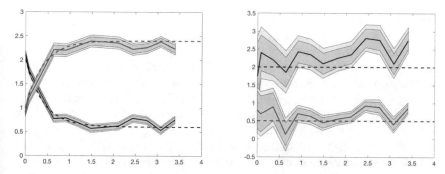

Fig. 13.2 Left panel: 75% (darker) and 90% (lighter) credibility envelopes of the concentrations. The values between observation points are plotted by linear interpolation of the credibility intervals. Right panel: The corresponding envelopes of the parameters k_1 and k_2. The generative values are indicated by the dashed curves

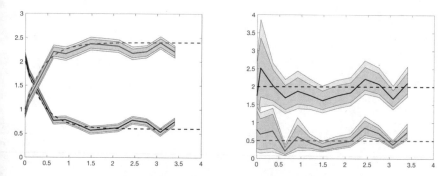

Fig. 13.3 This figure shows the same quantities as Fig. 13.2, when the correction step is computed by using logarithms of k_1 and k_2 as particle components to avoid negative values

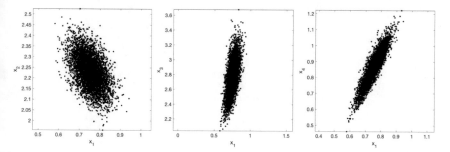

Fig. 13.4 Scatter plot of x_1^ℓ versus x_k^ℓ with $k = 2$ (left), $k = 3$ (center) and $k = 4$ (right), corresponding to the last time instance. The parameter values x_k, $2 \leq k \leq 4$, are based on the correlation with the observed variable x_1

Extensions of Kalman filtering include the Extended Kalman filtering (EKF), where non-linearities are locally linearized, and the Kalman filtering is used iteratively. The methodology is extensively used in navigation and GPS. Ensemble Kalman filtering was developed in the framework of data assimilation, with applications to meteorology and oceanography where the forward models tend to be so complex that large particle ensembles are impossible to use [31]. The concept of Randomize, then optimize (RTO) that we used as a motivation for the analysis step in EnKF goes back to the article [2]. An alternative approach combining ideas from linear filtering and particle filtering to reduce the ensemble size is the Unscented Kalman filtering (UKF), where the covariance in the analysis step is approximated by propagating a small ensemble of cleverly chosen particles that in the Gaussian case would be sufficient for the full recovery of the posterior covariance matrix [47, 48].

References

1. Anderson DF, Seppäläinen T and Valkó B (2017) *Introduction to probability*. Cambridge University Press
2. Bardsley JM, Solonen A, Haario H and Laine M (2014) Randomize-then-optimize: A method for sampling from posterior distributions in nonlinear inverse problems. SIAM Journal on Scientific Computing **36**: A1895–A1910
3. Bayes T (1763) An Essay towards solving a Problem in the Doctrine of Chances. Philosophical Transactions of the Royal Society of London **53**: 370–418
4. Berger J (2006) The case for objective Bayesian analysis. Bayesian Analysis **1**: 385–402
5. Billingsley P (2012) *Probability and measure.* Anniversary edition, John Wiley & Sons
6. Brooks S, Gelman A, Jones G and Meng XL (eds.) (2011) *Handbook of Markov chain Monte Carlo.* CRC press
7. Buhmann MD (2003) *Radial basis functions: theory and implementations.* Cambridge University Press
8. Calvetti D (2007) Preconditioned iterative methods for linear discrete ill-posed problems from a Bayesian inversion perspective. Journal of Computational and Applied Mathematics **198**: 378–395
9. Calvetti D, Hakula H, Pursiainen S and Somersalo E (2009) Conditionally Gaussian hypermodels for cerebral source localization. SIAM Journal on Imaging Science **2**: 879–909
10. Calvetti D, Kaipio JP and Somersalo E (2006) Aristotelian prior boundary conditions. International Journal of Mathematics and Computer Science **1**: 63–81
11. Calvetti D, Pascarella A, Pitolli F, Somersalo E and Vantaggi B (2015) A hierarchical Krylov-Bayes iterative inverse solver for MEG with physiological preconditioning. Inverse Problems **31**: 125005
12. Calvetti, D, Pascarella A, Pitolli F, Somersalo E. and Vantaggi B (2019) Brain Activity Mapping from MEG Data via a Hierarchical Bayesian Algorithm with Automatic Depth Weighting. Brain Topography **32**: 363–393
13. Calvetti D, Pitolli F, Somersalo E and Vantaggi B (2018) Bayes meets Krylov: Statistically inspired preconditioners for CGLS. SIAM Review **60**: 429–461
14. Calvetti D, Pragiola M, Somersalo E and Strang A (2020). Sparse reconstructions from few noisy data: analysis of hierarchical Bayesian models with generalized gamma hyperpriors. Inverse Problems **36**: 025010
15. Calvetti D, Pragliola M and Somersalo E (2020) Sparsity promoting hybrid solvers for hierarchical Bayesian inverse problems. SIAM Journal on Scientific Computing **42**: A3761–A3784

© The Editor(s) (if applicable) and The Author(s), under exclusive license to Springer
Nature Switzerland AG 2023
D. Calvetti and E. Somersalo, *Bayesian Scientific Computing*, Applied Mathematical
Sciences 215, https://doi.org/10.1007/978-3-031-23824-6

16. Calvetti D and Somersalo E (2005) Statistical elimination of boundary artefacts in image deblurring. Inverse Problems **21**: 1697
17. Calvetti D and Somersalo E (2005) Priorconditioners for linear systems. Inverse problems **21**: 1397
18. Calvetti D and Somersalo E (2007) *Introduction to Bayesian Scientific Computing – Ten Lectures on Subjective Computing.* Springer Verlag, New York
19. Calvetti D and Somersalo E (2008) Hypermodels in the Bayesian imaging framework. Inverse Problems **24**: 034013
20. Calvetti D and Somersalo E (2018) Inverse problems: From regularization to Bayesian inference. Wiley Interdisciplinary Reviews: Computational Statistics **10**: e1427
21. Calvetti D and Somersalo E (2022) *The less is more linear algebra of vector spaces and matrices.* SIAM, Philadelphia.
22. Calvetti D, Somersalo E and Strang A (2019) Hierarchical Bayesian models and sparsity: ℓ_2-magic. Inverse Problems **35**: 035003 (26pp)
23. Candès EJ, Romberg J and Tao T (2006) Robust uncertainty principles: Exact signal reconstruction from highly incomplete frequency information. IEEE Transactions on Information Theory **52**: 489–509
24. Chen SS, Donoho DL and Saunders MA (2001) Atomic decomposition by basis pursuit. SIAM Rev **43**: 129–159
25. Cotter SL, Roberts GO, Stuart AM and White D (2013) MCMC methods for functions: modifying old algorithms to make them faster. Statistical Science **28**: 424–446
26. De Finetti B (2017) *Theory of probability: A critical introductory treatment* (Vol. 6). John Wiley & Sons
27. Donoho DL (2006) Compressed sensing. IEEE Transactions on Information Theory **52**: 1289–1306
28. Doucet A and Johansen AM (2009) A tutorial on particle filtering and smoothing: Fifteen years later. Handbook of nonlinear filtering, **12**: 656–704
29. Doucet A, De Freitas N and Gordon NJ (2001) *Sequential Monte Carlo methods in practice* (Vol. 1, No. 2). Springer, New York
30. Duane S, Kennedy AD, Pendleton BJ and Roweth D (1987) Hybrid Monte Carlo. Phys letters B **195**: 216–222
31. Evensen G (2009) *Data assimilation: the ensemble Kalman filter* (Vol. 2) Springer, Berlin
32. Fletcher R (1987) *Practical methods of optimization* (2nd ed.) John Wiley & Sons, New York
33. Gelfand AE and Smith AFM (1990) Sampling based approaches to calculating marginal densities. Journal of the American Statistical Association **85**:398–409
34. Geman S and Geman D (1984) Stochastic relaxation, Gibbs distributions and the Bayesian restoration of images. IEEE Transactions on Pattern Analalysis and Machine Intelligence **6**:721–741
35. Ghahramani S (1996) *Fundamentals of Probability.* Prentice Hall
36. Gikhman II and Skorokhod AV (2004 The theory of stochastic processes (Reprint of the 1974 edition). Springer Science & Business Media.
37. Marques EC, Maciel N, Naviner L, Cai H and Yang J (2018) A review of sparse recovery algorithms. IEEE access **7**: 1300–1322
38. Gilks WR, Richardson S and Spiegelhalter DJ (1996) Markov Chain Monte Carlo in Practice. Chapmann & Hall
39. Haario H, Saksman E and Tamminen J (2001) An adaptive Metropolis algorithm. Bernoulli **7** 223–242
40. Hairer M, Stuart AM and Vollmer SJ (2014) Spectral gaps for a Metropolis-Hastings algorithm in infinite dimensions. The Annals of Applied Probability **24**: 2455–2490
41. Hanke M and Hansen PC (1993) Regularization methods for large-scale problems. Surveys in Mathematics in Industry **3**: 253–315
42. Hansen PC (2013) Oblique projections and standard-form transformations for discrete inverse problems. Numerical Linear Algebra with Applications **20**: 250–258

43. Hastings WK (1970) Monte Carlo sampling methods using Markov chains and their applications. Biometrika **57**: 97–109
44. Hestenes MR and Stiefel E (1952) Methods of Conjugate Gradients for Solving. Journal of research of the National Bureau of Standards **49**: 409–436
45. Hoffmann K and Kunze RA (1971) *Linear algebra*. Prentice-Hall, New Jersey
46. Jeffrey R (2004) *Subjective probability: The real thing*. Cambridge University Press
47. Julier SJ and Uhlmann JK (1997) New extension of the Kalman filter to nonlinear systems. Signal processing, sensor fusion, and target recognition VI **3068**: 182–193
48. Julier SJ and Uhlmann JK (2004) Unscented filtering and nonlinear estimation. Proceedings of the IEEE **92**: 401–422
49. Kaipio J and Somersalo E (2004) *Statistical and Computational Inverse Problems*. Springer Verlag, New York
50. Kalman RE (1960) A new approach to linear filtering and prediction problems. ASME Journal of Basic Engineering **82D**: 35–45
51. Kimeldorf GS and Wahba G (1970) A correspondence between Bayesian estimation on stochastic processes and smoothing by splines. The Annals of Mathematical Statistics **41**: 495–502
52. Liu JS (2003) *Monte Carlo strategies in scientific computing*. Springer NewYork Berlin Heidelberg
53. Liu J and West M (2001) Combined parameter and state estimation in simulation-based filtering. In: *Sequential Monte Carlo methods in practice* (pp. 197–223), Springer, New York
54. Matérn B (1960) Spatial variation. Meddelanden frøan Statens Skogsforskrignsinstitut **49**: 1–144
55. Metropolis N, Rosenbluth AW, Rosenbluth MN, Teller AH and Teller E (1953) Equations of state calculations by fast computing machine. Journal of Chemical Physics **21**: 1087–1091
56. Meyer CD (2000) *Matrix analysis and applied linear algebra*. SIAM, Philadelphia
57. Natterer F (2001) *The mathematics of computerized tomography* Society for Industrial and Applied Mathematics, Philadelphia
58. Neal RM (2011) MCMC using Hamiltonian dynamics. In: Brooks et al. *Handbook of Markov chain Monte Carlo*
59. Nummelin E (2002) MC's for MCMC'ists. International Statistical Review **70**: 215–240
60. Persson PO and Strang G (2004) A simple mesh generator in MATLAB. SIAM Review **46**: 329–345
61. Pragliola M, Calvetti D and Somersalo E (2022) Overcomplete representation in a hierarchical Bayesian framework. Inverse Problems and Imaging **16**: 19–38
62. Riesz F and Sz.-Nagy B (1990) *Functional analysis*. Dover, New York
63. Ristic B, Arulampalam S and Gordon N (2004) *Beyond the Kalman filter*. Artech House, Boston-London
64. Roberts GO and Tweedie RL (1996) Exponential convergence of Langevin distributions and their discrete approximations. Bernoulli **2**: 341–363
65. Rossi RJ (2018) *Mathematical Statistics : An Introduction to Likelihood Based Inference*. John Wiley & Sons, New York
66. Roininen L, Huttunen J M and Lasanen S (2014). Whittle-Matérn priors for Bayesian statistical inversion with applications in electrical impedance tomography. Inverse Problems Imaging **8**: 561–586
67. Roininen L, Lehtinen M, Lasanen S, Orispää M and Markkanen M (2011) Correlation priors. Inverse Problems and Imaging **5**: 167–184
68. Rudin W (1987) *Real and complex analysis (3rd edition)*. McGraw-Hill, New York
69. Saad Y (2003) *Iterative methods for sparse linear systems*. SIAM, Philadelphia
70. Stigler SM (1986) Laplace's 1774 Memoir on Inverse Probability, Statistical Science **1**: 359–363
71. Strang G (2006) *Linear algebra and its applications* Thomson, Brooks/ColemBelmont, CA Strang G (1993) *Introduction to linear algebra* . Wellesley-Cambridge Press, Wellesley, MA
72. Bulirsch R and Stoer, J (2002) *Introduction to numerical analysis (Vol. 3)*. Springer Verlag, Heidelberg

73. Stratonovich RL (1960) Application of the Markov processes theory to optimal filtering. Radio Engineering and Electronic Physics **5**: 1–19

74. Tarantola A (2005) *Inverse problem theory and methods for model parameter estimation.* SIAM, Philadelphia

75. Tarantola A and Valette B (1982) Inverse problems = quest for information, Journal of Geophysics **50**: 159–170

76. Tibshirani R (1996) Regression shrinkage and selection via the lasso. Journal of the Royal Statistical Society: Series B (Methodological) **58** 267–288

77. Tihonov AN (1963) On the solution of ill-posed problems and the method of regularization, Doklady Akademii Nauk SSSR **151**: 501–504 (Russian)

78. Tihonov AN (1963) On the regularization of ill-posed problems. Doklady Akademii Nauk SSSR textbf153: 49–52 (Russian)

79. Tikhonov AN and Arsenin VY (1977) *Solutions of ill-posed problems.* VH Winston & Sons

80. Trefethen LN and Bau III D (1997) *Numerical linear algebra.* SIAM, Philadelphia

81. Van Loan CF (1976) Generalizing the singular value decomposition, SIAM Journal on Numerical Analysis **13**: 76–83

82. West M (1993) Approximating posterior distributions by mixtures. Journal of the Royal Statistical Society **55**: 409–422

83. Whittle P (1954) On stationary processes in the plane. Biometrika **41**: 434–449

Index

© The Editor(s) (if applicable) and The Author(s), under exclusive license to Springer 283
Nature Switzerland AG 2023
D. Calvetti and E. Somersalo, *Bayesian Scientific Computing*, Applied Mathematical
Sciences 215, https://doi.org/10.1007/978-3-031-23824-6